HAWKS
FROM EVERY ANGLE

HAWKS
FROM EVERY ANGLE

HOW TO IDENTIFY RAPTORS IN FLIGHT

JERRY LIGUORI

Foreword by
DAVID A. SIBLEY

PRINCETON UNIVERSITY PRESS   PRINCETON AND OXFORD

Published by Princeton University Press, 41 William Street, Princeton, New Jersey 08540
In the United Kingdom: Princeton University Press, 3 Market Place, Woodstock, Oxfordshire OX20 1SY

Library of Congress Cataloging-in-Publication Data
Liguori, Jerry, 1966-
Hawks from every angle: how to identify raptors in flight / text and photos by Jerry Liguori ; foreword by David A. Sibley.
p. cm.
Includes bibliographical references and index.
ISBN-13: 978-0-691-11824-6 (cloth : alk. paper) ISBN-10: 0-691-11824-8 (cloth : alk. paper)
ISBN-13: 978-0-691-11825-3 (pbk. : alk. paper) ISBN-10: 0-691-11825-6 (pbk. : alk. paper)
1. Falconiformes—North America—Identification. 2. Falconiformes—Flight—North America—Pictorial works. I. Title.
QL696.F3L54 2005
598.9′44—dc22 2004066026

British Library Cataloging-in-Publication Data is available
This book has been composed in Sabon with Gill Sans Family Display
Printed on acid-free paper. ∞

nathist.princeton.edu

Composition by Bytheway Publishing Services
Printed in Italy by Eurographica

10 9 8 7 6 5 4 3 2 1

To my wife, Sherry, whose love and support make the pursuit of my dreams and goals possible

Contents

Foreword by David A. Sibley ix
Preface xi
Acknowledgments xi
Introduction 1
How to Use This Guide 1
Glossary 1
Anatomy 3
Flight Positions 4
Light Conditions 5
Understanding Molt 5
Aberrant Plumages 6
Hawk Migration 7
Migration Sites 7
Weather 10
Optics for Hawkwatching 10
Photography and Digital Imaging 10

Accipiters 15
Sharp-shinned Hawk 19
Cooper's Hawk 22
Northern Goshawk 24
Accipiter Shapes 29

Northern Harrier 31
Northern Harrier Shapes 39

Buteos 41
Red-shouldered Hawk 43
Broad-winged Hawk 46
Swainson's Hawk 49
Red-tailed Hawk 53
Ferruginous Hawk 67
Rough-legged Hawk 72
Buteo Shapes 83

Falcons 85
American Kestrel 86
Merlin 88
Peregrine Falcon 94
Prairie Falcon 98
Similar Species 99
Gyrfalcon 99
Mississippi Kite 100
Falcon Shapes 105

Vultures, Osprey, Eagles 107
Black Vulture 108
Turkey Vulture 109
Similar Species 109
Zone-tailed Hawk 109
Osprey 112
Bald Eagle 114
Golden Eagle 120
Osprey, Eagle, and Vulture Shapes 130

Bibliography 131
Index 133

Foreword

Watching hawks is a particular type of birdwatching; it could be called a method, a style, or a technique. If you want to observe hawks, you can just roam around and take your chances, or you can go "hawkwatching." This involves traveling to sites where migrating raptors concentrate—windswept ridges or mountaintops, the tips of coastal peninsulas, or the tops of waterside cliffs. In other words, go to a scenic place with a commanding view and sit for a while, scanning the open sky for hawks. If birds are not visible with the naked eye, scan for distant birds using your binoculars, and then try to identify the flickering gray specks that you discover.

Hawks are sometimes seen at close range, but as a rule hawk identification involves distance. Either way you are forced to identify birds using clues such as wing shape, wing posture, tail shape and proportions, wing beats, or other subtleties that may be unfamiliar to the backyard birder. It is out of necessity that we identify hawks this way. Hawks, by nature, are scarce and wary. In flight, small songbirds at a quarter mile are simply ignored by most birders—we'll wait to see others at close range when they are more easily identified. But without identifying the raptors we see at a quarter mile, we would identify very few. It follows then that hawk identification techniques have evolved to focus on clues that can be seen at long range, but once learned these same methods can be applied to any birds. Identifying small songbirds in flight at a quarter mile is possible if you truly desire to do so.

Early hawkwatchers who tried to identify raptors in flight based on the plumage differences they observed on stuffed specimens understandably made slow progress. This prompted some experts to suggest that distinguishing Cooper's from Sharp-shinned Hawks in the field would never be possible. Roger Tory Peterson's first field guide, published in 1927, included only one illustration of each species of hawk in flight. The illustrations were stiff and stylized, designed only to show a few key feather markings and only general differences in shape or posture. Even Peterson's 1980 revised edition suggests that Cooper's and Sharp-shinned Hawks can be distinguished only by size or tail shape (neither of which, we know now, is the best way to tell the two species apart).

When I began hawkwatching in the mid-1970s, identification was a mysterious and inaccessible skill. Since there were no published guides that went beyond the basics, one's knowledge could only be gained through long personal experience. Modern hawkwatching flourished in the 1980s and has advanced rapidly since then. Hawk identification now focuses not on a few clear-cut field marks but on an accumulation of subtle traits such as general color tones, wing and tail shapes, wing postures, overall proportions, manner of flight, and countless other subjective features.

There is an art and a science to identifying hawks. Sometimes well-defined, easily described field marks are obvious. Other times it is necessary to know technical details such as the age of a bird or the difference between old and new feathers. Hawk identification often involves judging whether a bird's wing beats are "choppy" or "snappy," whether its wing tips droop in a particular way, or an even more vague and indefinite sense of how a bird moves.

Through long hours in the field, hawk experts have been working out these differences—mostly by watching hawks pass by over and over again, until eventually identification becomes intuitive. A "speck" can be seen in the distance, and some indescribable characteristic regarding its movements will bring a name to the mind of the experienced hawkwatcher. The next step, which few people ever have the curiosity or the determination to ask, let alone answer, is the question "How did I know that was going to be a . . . ?" The answer requires a thoughtful and analytical approach to try to dissect the process, make hypotheses, test and revise, and understand, until eventually one can say with confidence, for example,

"Since goshawks have the longest hands of the accipiters, they have the most pronounced droop to their wings" (p. 27).

Jerry Liguori has spent most of the last twenty years in the field watching and photographing hawks, and thousands of hours poring over photos and research to piece together the puzzle of identification. The result of his patience and determination is this guide, which is the most detailed and confident explanation yet of the myriad clues that lead to successful identification of hawks.

This book is the first of its kind that deals with the real-world problems of identifying flying raptors from different angles: head-on, wing-on (from the side), tail-on (from behind), as well as the traditional "from below." It explains not only what to look for but also what not to look for, and when certain field marks are *not* visible. The understanding of what hawkwatchers actually face in the field comes through on every page of this book. Jerry Liguori has been there, and we are fortunate to have some of his experience now in book form.

David A. Sibley

Preface

My first visit to Hawk Mountain, Pennsylvania, was truly enlightening. I was familiar with identifying raptors at coastal sites but was intrigued at how certain species varied in appearance at a ridge site. For the first time, I observed hawks at eye level and below, and as I struggled to name each raptor that passed, I realized I was not as skilled at identifying hawks as I had thought. Since that day, I have come to understand that many factors influence the appearance of a raptor in flight, and that the art of identification is an ongoing learning process.

At one time or another I have mistaken almost every raptor species for something else. These mistakes most likely occurred because of a lack of experience, and the more time I spent studying hawks, the more familiar I became with them. I was fortunate to spend time watching hawks with experienced observers at a variety of migration sites. I learned to combine overall shape (silhouette), flight style (manner in which a bird flies), and plumage (coloration) to make accurate field identifications. Over time, I developed an approach to raptor identification that keys in on the distinguishing features of each species from various angles at which they are viewed (i.e., soaring, head-on, gliding overhead, wing-on, going away). This method has helped make identification more straightforward and accurate for me, and when I have taught it to others I have received encouraging feedback. This guide is a culmination of my experiences studying and teaching flight identification of raptors, and I hope it makes raptor identification easier and more fun for you.

ACKNOWLEDGMENTS

I thank all those who influenced me or encouraged me to complete this book, especially my wife, Sherry, who reviewed the entire book and helped in every phase of its completion. I also especially want to thank John Rokita, whose love of birds is a lasting inspiration to me, and Brian Sullivan, both of whom reviewed the entire text and made countless needed suggestions. Thanks also to Sherry's and my family for their support and encouragement. Vic Berardi, Jamie Cameron, Pete Gustas, and Mark Vekasy also reviewed parts of this guide. Thanks also go to Dave Andersen, Tom Aversa, Don Baccus, Aaron "Skippy" Barna, Andy Bauch, Marina Bean, Doreen Beiber, Mike Bisignano, Cress Bohnn, Ryan Brady, Claudia Burgos, Tom Carrolan, Efrain Castillejos, Bill Clark, Jack Conner, Claire Crow, Louie Dombroski, Kara Donohue, Keith Dowling, Pete Dunne, Charles Eldemire, Vince Elia, Brett Ewald, Mike Fitzpatrick, Ted Floyd, Tricia Franz, Sarah Frey, Wendy Gibble, Matt Gilligan, Liza Gray, Mike Green, Jason Guerard, Tanya Hashorva, Sue Hopkins, Marc Horowitz, Julian Hough, Kendall Jenkins, Eric Jepsen, Dimitri Karetnikov, Kevin Karlson, Paul Kerlinger, Robert Kirk, Helena Kokes, Ulf Konig, Mike Lanzone, Jon Larabee, Sheila Lego, Paul Lehman, Tony Leukering, Michelle Lockwood, Josh Lowery, Dylan Maddox, Rob Mahedy, Joe Mangino, Brian Maransky, Mark McCaustland, Kirsten McDonnell, Nathan McNett, Brian Meiering, Lester Melendez, Rigoberto Mendoza, Doug Merkle, Susie Michaelson, Pam Mikula, Trish Miller, Jeannette Morss, Nicole Munkwitz, Marleen Murgitroyde, Chris Neri, Frank and Kate Nicoletti, Dan Niven, Michael O'Brien and Louise Zemaitis, Marti Ouelette, James Paolino, Chris Peper, Rod and Marleen Planck, Kat Poetter, Fernando Rincon, Andre Robinson, Rafael Rodriguez, Dale Rosselet, Greg Ryder, Deneb Sandack, Jeff Schultz, Bill Seng, Michael Shupe, David Sibley, Jeff Smith, Larissa Smith, Ruth Smith, Zach Smith, Eric and Lydia Stiles, Jon Jon Stravers, Marty Stum, Clay and Pat Sutton, Dave and Kathy Tetlow, Sarah Thomsen, Matt Tribulski, Tono Urona, Lena Usyk, Chris Wagner, Joan Walsh, Dick Walton, Jim Watson, Patrick Watson and Joy Emory, Aimee Weldon, Brian Wheeler, Stephen Wilson, the Cape May Raptor Banding Project of 1989, 1993, 1994, 1995, 1996, and 1997, the Braddock Bay Raptor Research staff of 1994, 1996, and 1997, the Goshutes crew of 1998, 1999, 2000, and 2001, Cape May Bird Observatory, HawkWatch International, Borge B. Andersen & Associates, and anyone else who shared in my experience learning about and watching birds.

Introduction

This book is designed to help birders of all skill levels identify raptors in flight—from those who want to name distant raptor silhouettes to those who want to identify a hawk chasing songbirds at their backyard bird feeder. The photos and text simplify raptor identification by illustrating the most practical and pertinent characteristics of shape, plumage, and flight style that distinguish each species. Field marks that are often obscure in flight, such as the wavy tail bands of juvenile Northern Goshawks, are not discussed. Note that watching birds in flight is very different from looking at photographs. Certain traits that may be obvious in photos, such as eye color, are often impossible to observe on flying birds. The distinct color tones or patterns exhibited by a bird's overall plumage, such as the two-toned appearance of Swainson's Hawks, are often more useful in identification than specific plumage markings and are presented as they normally appear in the field.

The 22 raptor species covered in this book are those that most commonly occur at migration sites throughout the United States and Canada. Some species herein also occur throughout Mexico, Central America, and South America, but these areas are not discussed in regard to range. Vultures, which are no longer classified as raptors, are addressed because they are often confused with hawks and eagles and are tallied at most migration sites. Species that are rarely observed on migration, or that are geographically localized, such as kites, Common Black-Hawk, Harris's Hawk, Gray Hawk, Short-tailed Hawk, White-tailed Hawk, Crested Caracara, and Gyrfalcon, are mentioned only briefly or are excluded, to lend space for difficult identification problems of the more commonly observed species. Particular races of certain species that are nonmigratory or rare at hawk migration sites, such as the Fuertes and Florida races of the Red-tailed Hawk, the California race of the Red-shouldered Hawk, and the Black race of the Merlin, are not discussed in detail. Familiarity with the common species is helpful when faced with identification of uncommon species.

HOW TO USE THIS GUIDE

This is primarily a visual guide; the photos and captions are the crux of the book and are meant to stand on their own. The style of presentation is unique in that species similar in appearance are shown alongside each other for simple comparison and quick reference. For example, the Turkey Vulture and Zone-tailed Hawk are taxonomically separate but are presented in direct comparison with each other because of their similarities in appearance. Since there are only three species of North American accipiters, which are extremely similar in appearance, the text for the accipiters is organized by age class for easy comparison.

The color photos were chosen to present plumage traits only, whereas the black-and-white photos depict shape characteristics. On the black-and-white pages, species of different sizes are all shown as the same, in order to stress shape features. This is also done on some of the color plates, in these cases to emphasize plumage differences. Color photos of birds "at a distance" are meant to represent how they often appear in the field.

GLOSSARY

Adult plumage Definitive or final plumage acquired.

Albinism Presence of some (partial albinism) or all (complete albinism) white feathers on a normally darker-plumaged bird (see Aberrant Plumages).

Auriculars Feathers on the sides of the head covering and surrounding the ears (see Anatomy).

Axillaries Wing pits (see Anatomy).

Bib Patch of dark feathers on the upper breast that contrasts with a paler body.

Buff-colored Light tan color.

Carpal Underwing area at the "wrist" where all the primaries meet (see Anatomy).

Cere Flesh between the bill and forehead (see Anatomy).

Dihedral Flight position in which wings are held above the plane of the body.
Dilute plumage Overall light tan plumage occurring in normally dark birds.
Flight feathers Includes primaries, secondaries, and tail feathers.
Glide To move forward with wings pulled in (see Flight positions).
Hand Part of the wing that consists of all the primaries.
Head-on Eye-level, front profile view (see Flight positions).
Hover To remain stationary in flight while flapping (see Flight positions).
Immature All plumages other than adult.
Intergrade Offspring of two adults of different races or morphs.
Juvenile First plumage of a bird.
Leading edge of the wing. Front edge of the wing.
Leggings Feathers covering the legs (see Anatomy).
Melanism Presence of unusually dark feathers.
Modified dihedral Flight position in which wings are raised at the shoulders and level at the wrists.
Molt Replacement of feathers, usually occurring from April through September. Many birds retain various feathers from the previous year.
Morph Color form.
Mustache Dark lines down the sides of the face seen on many falcons.
Nape Back of neck (see Anatomy).
Patagium Skin between the wrist and body along the leading edge of the wing (see Anatomy).
Primaries Ten outermost remiges forming the "hand" of the wing; the outermost notched primaries make up the "fingers" of a hawk (see Anatomy).
Primary coverts Feathers covering the base of the primaries.
Primary projection Distance the primaries extend beyond the base of the wing. On passerines, the distance the longest primary extends beyond the secondaries while perched.
Remiges Secondaries and primaries.
Rufous Rusty or orange tint.
Scapulars Feathers along the sides of the back.
Secondaries Flight feathers from the wrist to the body, making up the "base" of the wing (see Anatomy).
Secondary coverts Feathers covering the base of the secondaries.
Soar To rise in a circular motion with wings outstretched (see Flight positions).
Stoop To dive from above with wings folded, usually in pursuit of prey (see Flight positions).
Sub-adult Age(s) between juvenile and adult.
Subterminal band Second-to-last band toward the tip of the tail.
Superciliary Pale feathers over the eye forming an eye-line (see Anatomy).
Terminal band Last band toward the tip of the tail or wings.
Trailing edge Back edge of the wing.
Undertail coverts Feathers covering the underside of the base of the tail (see Anatomy).
Underwing coverts Feathers covering the base of the underwing (see Anatomy).
Uppertail coverts Feathers covering the top of the base of the tail (see Anatomy).
Upperwing coverts Feathers covering the base of the upperwing (see Anatomy).
Wing-on Eye-level, side profile view (see Flight positions).
Wing panel Pale or translucent "window" throughout the primaries (see Anatomy).
Wrist Joint toward the middle of the leading edge of the wing.

ANATOMY

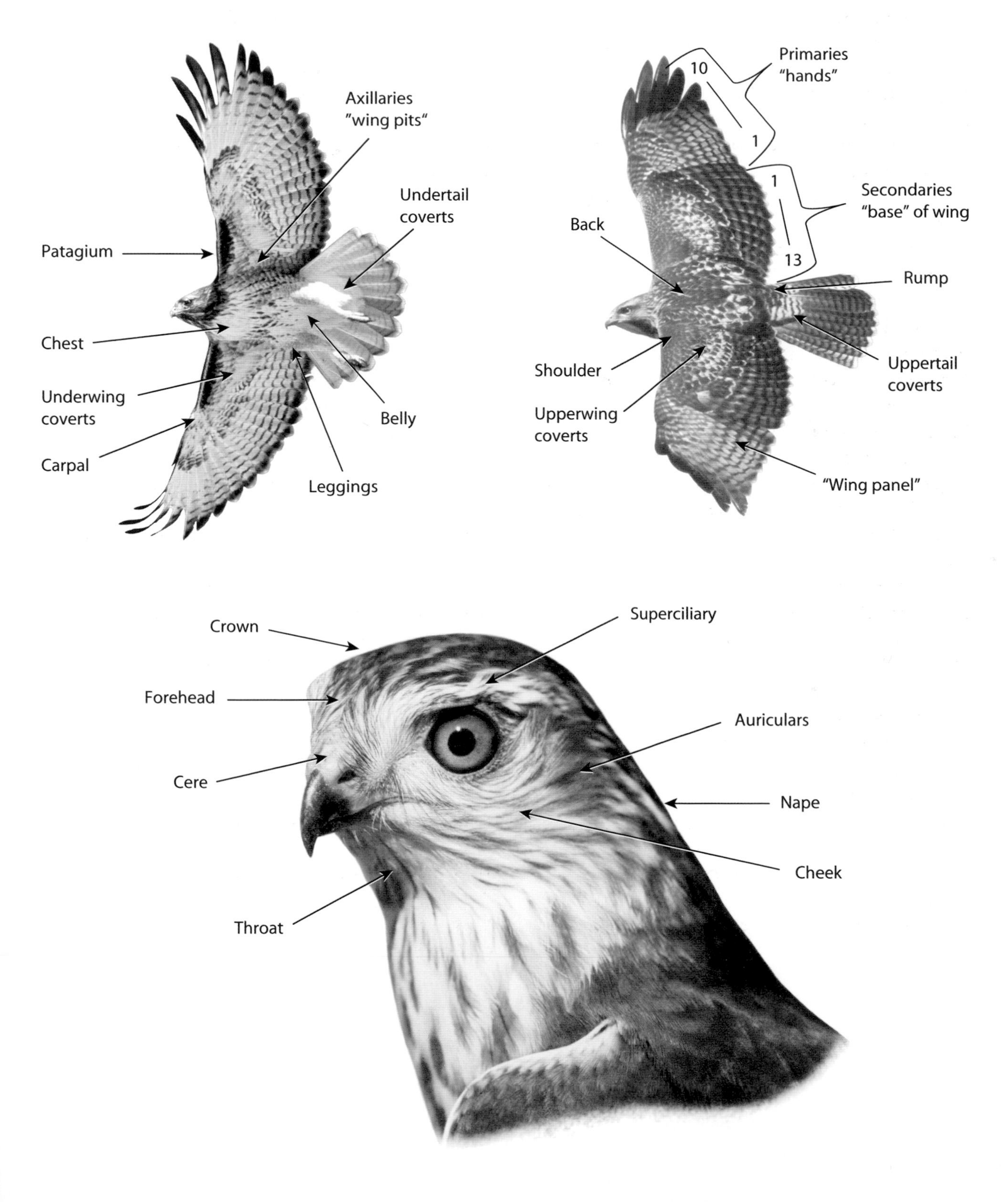

Axillaries
"wing pits"
Undertail
coverts
Patagium
Chest
Underwing
coverts
Carpal
Belly
Leggings
Primaries
"hands"
10
1
1
13
Secondaries
"base" of wing
Back
Rump
Shoulder
Uppertail
coverts
Upperwing
coverts
"Wing panel"
Crown
Superciliary
Forehead
Auriculars
Cere
Nape
Cheek
Throat

FLIGHT POSITIONS

Soaring

Head-On

Gliding

Wing-On

Going away

Flapping

Hovering

Stooping

LIGHT CONDITIONS

Understanding how various light conditions can affect the appearance of raptors is important in identification. Lightly colored birds can lack contrast and appear uniformly dark in poor light, such as when backlit or against a cloudy sky (see figure 1), whereas true dark morph birds show a contrast underneath between the body and flight feathers. By contrast, birds can look paler than usual when illuminated by highly reflective ground cover such as snow, sand, and pale grasses (see figure 2). The appearance of a bird's coloration can also be affected when the sun is low on the horizon. In this setting, an orange glow is cast to the underside (see figure 3), making identifying or aging raptors difficult in certain cases. Several conditions affect the way a bird's size and shape are perceived. Against a bright blue sky, birds often appear smaller than usual; however, birds may appear larger than usual when observed against cloud cover.

Figure 1 *Figure 2*

Figure 3

UNDERSTANDING MOLT

The extent to which an individual raptor molts each year varies according its size, age, health, and extent of migration. Smaller raptors, such as Sharp-shinned Hawks, usually undergo a complete molt during one cycle (late spring to late summer). Birds that have a relatively long migration and short breeding season, such as Swainson's Hawks, tend to take two years to complete a molt. Large birds, such as eagles, may take up to several years to complete a molt. In some cases, eagles of different ages may resemble each other because of the variations in their molt. Juvenile raptors, which are more likely to be food stressed than older birds, tend to migrate farther than adults and typically retain varying amounts of juvenile feathers after their first molt cycle.

The shape of a bird can be altered as a result of molt. For example, a bird with outer primaries missing or primaries that are partially grown in appears more pointed winged or blunt winged than usual. When tail feathers are in the process of molt, a bird may look shorter tailed than usual, making its overall shape appear atypical (see figure 4). When the central tail feathers are missing, a bird's tail may appear squared or notched at the tip when folded. Molt can also affect the appearance of a bird's plumage pattern. For example, each flight feather of an adult Golden Eagle has a white base, which is hidden by the underwing linings. During spring and summer, when the wing linings are undergoing replacement, the white base of the adult flight feathers may become visible, making the underwing pattern appear similar to that of immature Golden Eagles (see Golden Eagle).

In some cases, the presence of molt is the determin-

Figure 4. Swainson's Hawks typically show pointed wings and a relatively long tail; however, the wing tips of this individual are less pointed than usual and the tail appears extremely short because of molt.

ing factor in telling the age of a bird. For example, juvenile raptors do not begin to molt flight feathers until their first spring. Therefore, **raptors in active flight feather molt or showing signs of previous flight feather molt during fall migration are not juveniles.** Some immature eagles, especially white-bellied Bald Eagles, can only be aged based on overall molt patterns (see Bald Eagle). Be aware that eagles in active molt can look ragged because of missing or partially grown flight feathers and can be difficult to age. Birds may lose or break feathers for odd reasons, whereas molt is typically symmetrical on each wing and on each side of the tail.

ABERRANT PLUMAGES

Some raptor species exhibit significant variation in plumage, but aberrant plumages, including albinism, melanism, and dilute plumage, can occur in any species. Birds exhibiting aberrant plumages are rare but pose interesting identification challenges. Understanding normal plumage variation and the effects of aberrance in plumage will facilitate the identification of problem individuals.

Albinism is the most common aberrant plumage that occurs in raptors. Most albinistic birds exhibit several white wing, body, or tail feathers, or a combination of each. Albinism is often symmetrical, appearing as white patches or splotches in identical areas on each side of a bird. In raptors, albinism is most common in Red-tailed Hawks, occurring in about 1 in every 15,000 (see figure 5). Albinism is extremely rare in juvenile raptors; complete albinism is rare in all raptor species.

Melanism, the replacement of normally light feathers with darker feathers, is rare in raptors but is documented in several species including Osprey, Northern Harrier, and Northern Goshawk. Melanistic birds can be difficult to identify because many raptors exhibit true dark plumages. Remember, most dark buteos have contrasting pale flight feathers, whereas melanistic birds typically have darker flight feathers than normal.

The term "dilute plumage" is used to describe birds that are paler than usual, or tan colored overall.

Figure 6. A succession of north-to-south-leading ridges, such as the Goshute Mountains in Nevada, provides lift for migrating hawks.

Figure 5. Albinistic birds, such as this Red-tailed Hawk, often appear whiter on the upperside (inset) than on the underside.

Figure 7. During migration, many hawks congregate along shorelines, especially peninsulas such as Cape May Point, New Jersey. © Sherry Liguori

Diagnostic but often muted field marks are usually present on albinistic, melanistic, and dilute plumaged birds. Familiarity with typical plumages helps in the identification of aberrant plumaged birds. I once witnessed what appeared to be a white morph Gyrfalcon that turned out to be an albinistic Peregrine Falcon. The bird was first identified by its lack of dark wing tips, a trait shown by all Gyrfalcons. Sometimes shape and flight style characteristics are the only telling traits noticeable on birds that exhibit aberrant plumages.

HAWK MIGRATION

Migration occurs across all of North America as raptors wander to and from their breeding grounds each spring and fall. Several factors influence the migratory pathways of raptors. The most significant of these is geography. Mountainous sites, such as the Goshute Mountains in Nevada (see figure 6) and Hawk Mountain, Pennsylvania, are famous for attracting concentrations of migrating hawks. As wind strikes a ridge, it is deflected upward, providing birds with a current of air, or updraft, along the ridge for a more energy-efficient means of travel. Often, raptors will travel along these updrafts for miles without a noticeable beat of their wings. Under optimal conditions, a succession of ridges can provide a migrant raptor with miles of nearly effortless flight.

Significant concentrations of migrating birds also occur along the shorelines of large bodies of water, such as the Great Lakes, the Atlantic and Pacific Oceans, and the Gulf of Mexico. Since thermals, pockets of warm, rising air used by soaring birds to gain altitude, do not form over water, most hawks hesitate to cross oceans and large lakes. Instead, they fly over land where they use thermals, rest, and hunt for food. Birds with relatively long or tapered wings, such as Bald Eagles, Ospreys, kites, Northern Harriers, falcons, and Rough-legged Hawks, readily cross large bodies of water. Official hawk migration counts are conducted at many sites along shorelines in North America. Peninsula sites in particular, such as Cape May Point in New Jersey (see figure 7), Whitefish Point in Michigan, and Golden Gate State Park in California, are excellent places to witness raptor migration. Birds faced with a peninsula often "funnel" toward its tip where they converge into large groups before redirecting their route. Sites near water barriers generally concentrate greater numbers of raptors than ridge sites.

The majority of fall hawk migration in North America occurs between September and mid-November. Peak migration dates for smaller hawks, such as American Kestrels, and for long-distance migrants, such as Broad-winged Hawks, tend to be earlier than those of large raptors or short-distance migrants (see table 1 at end of Introduction). Migration in the West may peak up to two weeks earlier than in the East. Likewise, coastal and southern sites usually experience a later peak than inland and northern sites. The bulk of spring migration occurs from mid-March to early May, with larger raptors peaking earliest. Juvenile birds usually peak earlier in fall and later in spring than adult birds of the same species. Generally, Ospreys and falcons are observed in greater numbers at coastal sites, whereas inland sites record more buteos and eagles. However, many sites record a variety of species in significant numbers.

Migrants do stray at times, and some species show up regularly at sites outside their normal range. Birders in the East are alert for western hawks, such as Swainson's Hawks, that show up annually in eastern North America on migration. Broad-winged Hawks, once thought of as rare visitors to western North America, are recorded at most western hawk migration sites, and recent nesting sites have been located in western North America.

MIGRATION SITES

There are more than 1,000 known hawk migration sites throughout North America, and new ones are discovered each year as raptor watching increases in popularity (see figure 8). Space does not permit descriptions of each official site, but websites do exist for many of them. HMANA (Hawk Migration Association of North America) publishes spring and fall journals that contain data from all reporting raptor migration sites in North America, articles on raptor migration and identification, and information on raptor-related gatherings, conferences, and meetings.

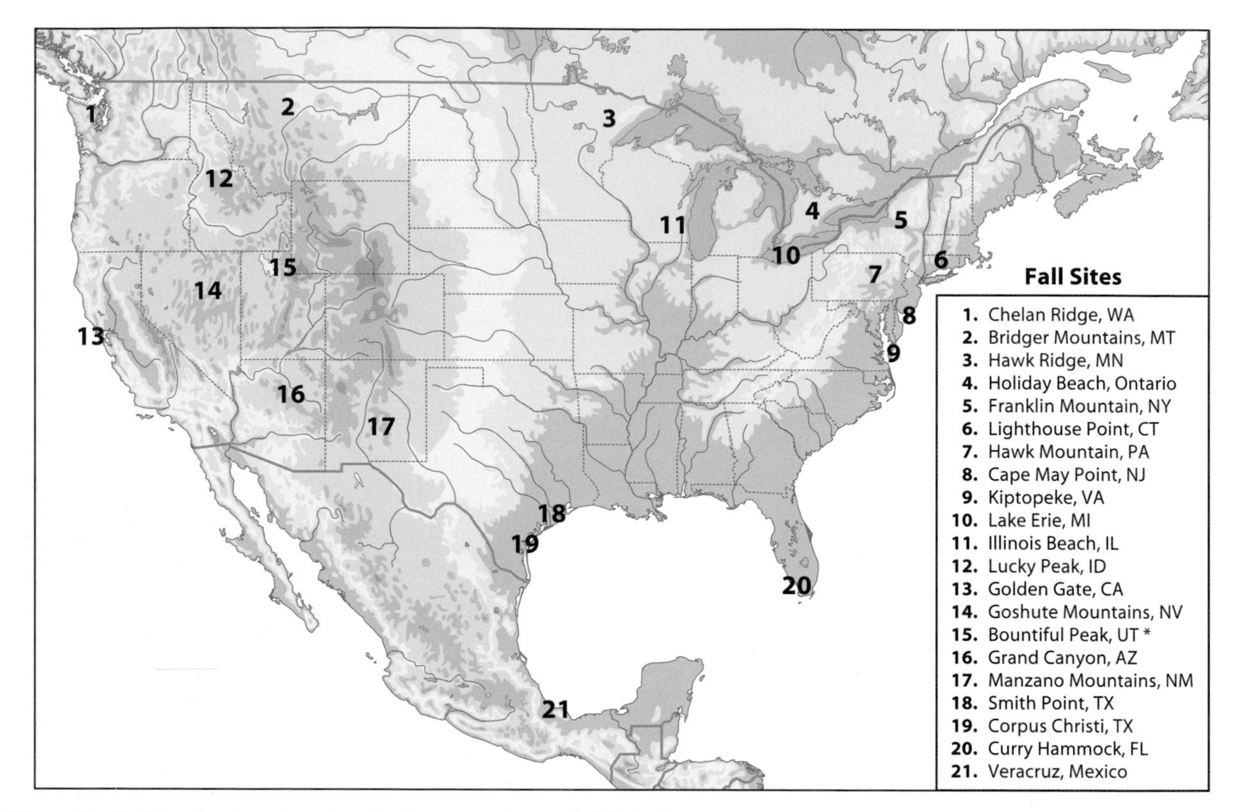

Figure 8a. Fall hawk migration sites (* denotes spring and fall site).

Cape May Point is one of the most popular sites in North America for watching hawks. Northwesterly winds throughout the fall push massive numbers of migrating birds east toward the Atlantic coast. Many raptors that reach the Atlantic coast north of Cape May follow the coastline south until they reach the Cape May Peninsula, where they either cross Delaware Bay or head north along the bayshore. Observers at Cape May record seven species of raptors (Osprey, Northern Harrier, Sharp-shinned Hawk, Cooper's Hawk, American Kestrel, Merlin, and Peregrine Falcon) in numbers equal to or greater than at any other site in North America (see table 2). Cape May Point an ideal place for birders to learn about birds and improve their hawk identification skills. It is not uncommon to see three species of falcon (American Kestrel, Merlin, and Peregrine Falcon) the same field of view at Cape May.

Northwest winds also produce the largest daily flights at the most renowned hawk migration site of all, Pennsylvania's Hawk Mountain. In 1934, Hawk Mountain Sanctuary conducted its first official autumn hawk count and has since become the most popular hawkwatching site in North America. With its beautiful scenery and spectacular views of migrant raptors, Hawk Mountain is a "must see" for birdwatchers. Mid-to late September is an ideal time to see Broad-winged Hawks and accipiters. Mid-October to mid-November is the peak time for larger birds such as Red-tailed and Red-shouldered Hawks, Northern Goshawks, and Bald and Golden Eagles.

With an astounding 360-degree view of mountains, valleys, and the Great Salt Desert, the Goshute Mountains in northeastern Nevada is one of my favorite sites to watch birds. Whereas most sites require specific wind conditions for peak flights, migrants concentrate during all conditions at the Goshutes; a steady wind is best for providing close-up views of raptors. This range is best known for its flights of Cooper's Hawks, tallying as many as 6,736 in a sin-

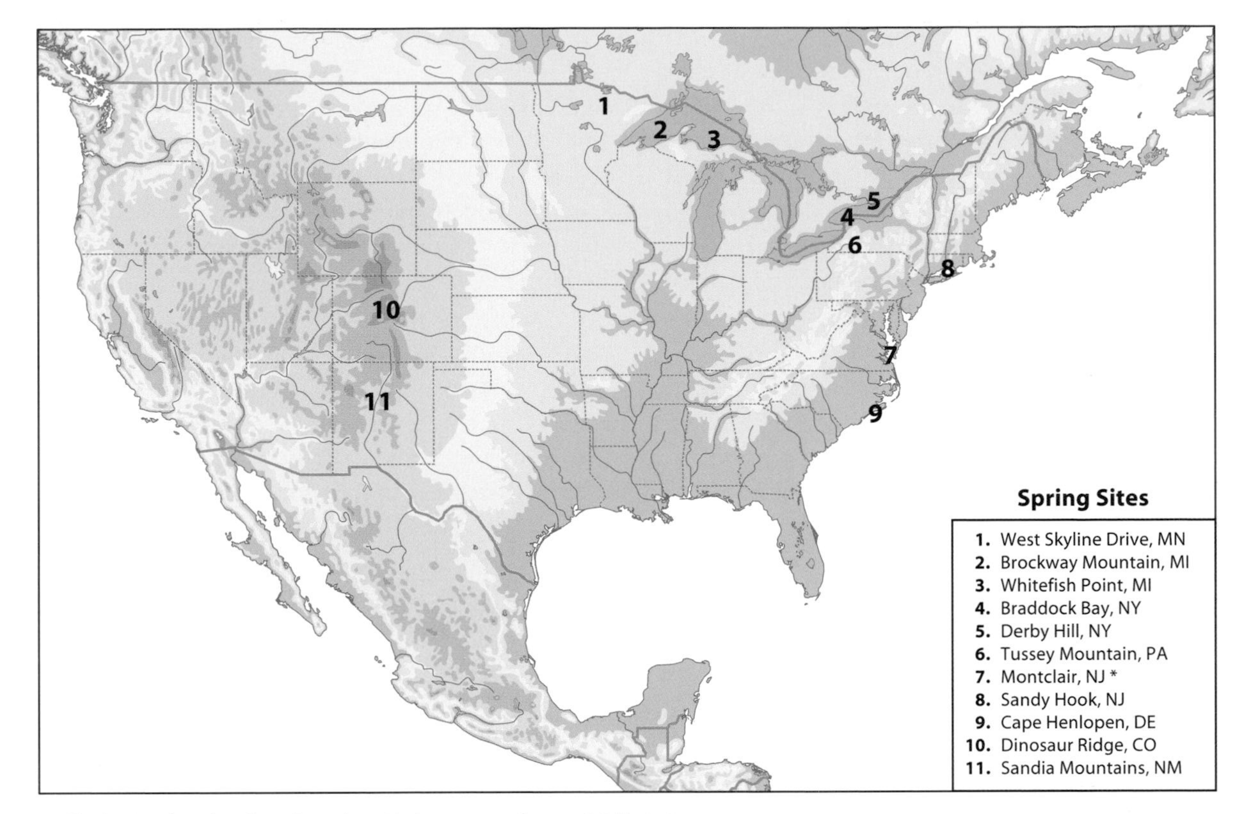

Figure 8b. Spring hawk migration sites (* denotes spring and fall site).

gle season (see table 3). It is also the most reliable site in the West to see Broad-winged Hawks, with as many as 160 sighted in one season. Dark morph Broad-winged Hawks, sought after by hawkwatchers, make up about 3 to 4 percent of the total Broad-winged Hawks in the West.

Hawkwatching along the Great Lakes is dynamic but can require patience while awaiting optimal weather conditions. With ideal weather, amazing one-day flights can occur, such as the 4,591 Red-tailed Hawks recorded at Derby Hill in April 1995 and the 1,414 Red-shouldered Hawks counted at Braddock Bay in March 1987. Braddock Bay and Derby Hill, located about 110 miles apart along the southern shore of Lake Ontario, record similar totals each spring. Official counters at each site have even seen several of the same birds on the same day! The majority of hawks seen at Derby Hill are in close proximity to the count site, whereas migrants at Braddock Bay are often distant. Harriers, accipiters, Rough-legged Hawks, and American Kestrels are also seen in excellent numbers along Lake Ontario's southern shoreline. Hawk Ridge along the southwestern shore of Lake Superior in Duluth, Minnesota, records fall totals similar to the spring totals at Braddock Bay and Derby Hill; however, Bald Eagles and Northern Goshawks are tallied in greater numbers at Hawk Ridge than at any other migration site. Impressive numbers of Sharp-shinned Hawks, American Kestrels, Red-tailed Hawks, and Rough-legged Hawks pass Hawk Ridge each fall as well.

No raptor book would be complete without mention of the site that hosts the largest migration of raptors in the world: Veracruz, Mexico, where millions of hawks and vultures are tallied each fall (see tables 2 and 3). The migration at Veracruz is so overwhelming at times that official observers liken the counting to estimating. It is believed that almost the entire North American populations of Broad-winged and Swainson's Hawks migrate over Veracruz each fall.

Corpus Christi, Texas, and Lake Erie Metropark, Michigan, are the only sites north of Veracruz that rival its numbers for Broad-winged Hawks. On September 17, 1999, official counters from Southern Michigan Raptor Research tallied an incredible 555,371 Broad-winged Hawks at Lake Erie Metropark. Corpus Christi reached a whopping seasonal count of 970,025 Broad-winged Hawks in fall 1998.

WEATHER

Most migration sites require specific weather conditions during certain time periods to produce peak hawk flights. Knowledge of the optimal weather conditions and dates for specific sites will increase one's chances of witnessing a big flight. Annual fluctuations for species such as Rough-legged Hawks and Northern Goshawks occur as well. Following these fluctuations is a good way to predict high or low counts the following season. For example, high counts of Rough-legged Hawks were recorded at numerous migration sites during fall 1999 and at wintering sites in 2000. Likewise, spring sites such as Whitefish Point tallied high counts of Rough-legged Hawks headed north in 2000.

During spring, many peak flights occur ahead of a warm front, as birds heading north use southerly prevailing winds. Most fall flights occur after the passage of a cold front, when northerly winds that assist birds heading south are most prevalent. Geography, however, determines which specific wind directions will lead birds to each site. The most favorable winds for ridge sites are those that strike the ridge at an angle that produces optimal lift. North-to-south ridges often experience good flights on various wind conditions. However, many ridge sites that do not run north to south may experience slow flights in less than optimal wind conditions. At coastal and shoreline sites, optimal winds are those that "push" birds toward the shorelines. Even during snow squalls or light drizzle, optimal wind conditions can produce significant hawk flights.

OPTICS FOR HAWKWATCHING

I strongly recommend that beginning birders equip themselves with high-quality binoculars. Be careful of compact binoculars, as some models offer a crisp image but have a small field of view, which can frustrate a birder who is not practiced at finding birds with binoculars. The magnification one chooses should be based on what is most comfortable. I use 7x (magnification) binoculars for watching hawks because I prefer an extrawide field of view; however, many birders prefer the added power of 8x to 10x binoculars.

A spotting scope can be useful for identifying hawks, but it may also hinder a beginning observer in several ways. It can be extremely difficult to locate birds in a spotting scope, especially in blue skies, because of the high magnification and small field of view. Moreover, a spotting scope is useless during periods of strong heat shimmer, which makes a scope unclear, or during gusty winds, when a scope is unstable. When using binoculars to identify birds, emphasis tends to be placed more on shape and flight style characteristics. When using a spotting scope, birders tend to focus mainly on plumage traits, which may be obscure under certain conditions. I have seen my own identification skills diminish after relying on a spotting scope for an extended period of time. In truth, skill level in identifying birds is more a matter of experience than of high-powered optics. Keep in mind that even with a great deal of practice and the finest optics, no one can identify every bird he or she sees!

PHOTOGRAPHY AND DIGITAL IMAGING

The majority of photographs in this guide were taken using a Canon EOS 3 body along with a Canon 300-mm/4.0 lens; a Canon 1.4x extender was used in most cases. My film of choice was Fuji Provia F ISO 100 pushed one stop to ISO 200. I began using a Canon 10D digital camera body along with my 300-mm lens in June 2003 and was able to acquire many photographs for this book in a short period of time. Since raptors will harass large owls that are conspicuous during the daytime, I used a plastic owl decoy at various migration sites to obtain close-up photos for plumage representation.

In some cases, a bird from one photograph was "pasted" into the background of another for side-by-side comparison. These comparison photographs are presented as plates in the manner that artwork is presented in most other field guides. In some cases, birds were inset into a frame to depict traits at a distance. Any manipulation of photographs in this guide was done to present identification traits in the most effective manner possible. Changes to the actual appearance of the birds were avoided so that the integrity of the photos was not compromised.

Table 1

Timetable of Raptor Migration

Species	Spring			Fall		
	March	April	May	Sept.	Oct.	Nov.
Black Vulture						
Turkey Vulture						
Osprey						
Bald Eagle						
Mississippi Kite						
Northern Harrier						
Sharp-shinned Hawk						
Cooper's Hawk						
Northern Goshawk						
Red-shouldered Hawk						
Broad-winged Hawk						
Swainson's Hawk						
Red-tailed Hawk						
Ferruginous Hawk						
Rough-legged Hawk						
Golden Eagle						
American Kestrel						
Merlin						
Peregrine Falcon						
Prairie Falcon						

Table 2

Daily High North American Migration Counts

Species	*Site*	*High Count*	*Date*
Black Vulture	Kiptopeke, VA	684	10/25/1995
Turkey Vulture	Veracruz, Mexico	707,798	10/17/2003
Osprey	Kiptopeke, VA	1,053	9/20/1996
Bald Eagle	West Skyline Drive, MN	822	3/24/2004
Mississippi Kite	Veracruz, Mexico	95,989	9/1/2002
Northern Harrier	Braddock Bay, NY	440	4/16/1996
Sharp-shinned Hawk	Cape May Point, NJ	11,096	10/4/1977
Cooper's Hawk	Goshute Mountains, NV	913	9/27/2001
Northern Goshawk	Hawk Ridge, MN	1,229	10/15/1982
Red-shouldered Hawk	Braddock Bay, NY	1,414	3/24/1987
Broad-winged Hawk	Veracruz, Mexico	775,760	9/28/2001
Swainson's Hawk	Veracruz, Mexico	782,653	10/17/2003
Red-tailed Hawk	Derby Hill, NY	4,591	4/11/1995
Ferruginous Hawk	Dinosaur Ridge, CO	40	3/10/1997
Rough-legged Hawk	Whitefish Point, MI	525	4/23/2000
Golden Eagle	South Livingstone, Alberta	1,071	10/8/2000
American Kestrel	Cape May Point, NJ	24,875	10/16/1970
Merlin	Cape May Point, NJ	867	9/30/1999
Peregrine Falcon	Curry Hammock, FL	521	10/1/2003
Prairie Falcon	Goshute Mountains, NV	7	9/2/1998

* This table contains the highest daily count for each raptor species tallied at North American migration sites.

Table 3

Seasonal High North American Migration Counts

Species	*Site*	*High Count*	*Date*
Black Vulture	Kiptopeke, VA	2,630	Fall 1996
Turkey Vulture	Veracruz, Mexico	2,677,355	Fall 2002
Osprey	Cape May Point, NJ	6,734	Fall 1996
Bald Eagle	Hawk Ridge, MN	4,368	Fall 1994
Mississippi Kite	Veracruz, Mexico	306,274	Fall 2002
Northern Harrier	Braddock Bay, NY	3,177	Spring 1996
Sharp-shinned Hawk	Cape May Point, NJ	61,167	Fall 1984
Cooper's Hawk	Goshute Mountains, NV	6,736	Fall 1998
Northern Goshawk	Hawk Ridge, MN	5,819	Fall 1982
Red-shouldered Hawk	Braddock Bay, NY	2,660	Spring 1990
Broad-winged Hawk	Veracruz, Mexico	2,389,232	Fall 2002
Swainson's Hawk	Veracruz, Mexico	1,197,850	Fall 2003
Red-tailed Hawk	Derby Hill, NY	19,531	Spring 1995
Ferruginous Hawk	Dinosaur Ridge, CO	241	Spring 1997
Rough-legged Hawk	Whitefish Point, MI	2,600	Spring 2000
Golden Eagle	Mt. Lorrette, Alberta	4,753	Fall 2000
American Kestrel	Cape May Point, NJ	21,821	Fall 1981
Merlin	Cape May Point, NJ	2,875	Fall 1985
Peregrine Falcon	Curry Hammock, FL	2,858	Fall 2003
Prairie Falcon	Sandia Mountains, NM	59	Spring 1998

* This table contains the highest seasonal count for each raptor species tallied at North American migration sites.

Accipiters

Sharp-shinned Hawk (*Accipiter striatus*), **Cooper's Hawk** (*Accipiter cooperii*), **Northern Goshawk** (*Accipiter gentilis*)

OVERVIEW

The three species of North American accipiters are similar in appearance. Although they vary in size, with Sharp-shinned Hawks the smallest and Northern Goshawks the largest, accipiters can be difficult to distinguish in flight even under ideal conditions. This is evident at most hawkwatching sites throughout North America, where "unidentified accipiters" make up the majority of unidentified raptors. Becoming familiar with the differences, especially in shape and flight style, between Sharp-shinned and Cooper's Hawks and between Cooper's Hawks and Northern Goshawks takes practice. Since Northern Goshawks are uncommon at most migration sites in North America, many birders are unfamiliar with them and find them difficult to identify. With repeated observations, the nuances of accipiter identification become familiar in the same way one recognizes a friend's silhouette or manner of walking.

Accipiters can soar almost as effortlessly as buteos and display short bursts of speed that rival that of falcons. Although all raptors may flap and glide intermittently, accipiters display this manner of flight most frequently. All three accipiter species occur throughout eastern and western North America.

Size and Structure

Accipiters inhabit woodlands and semiopen country, and their shapes have evolved to suit these environments. With relatively short wings and a long tail, they can maneuver with great ease through thick forest in pursuit of prey. Their wings are broad and slightly rounded at the tips, giving them an overall stocky appearance. Accipiters vary slightly in shape between adults and juveniles, and males and females. Adults have a shorter tail and slimmer wings than juveniles. Females have longer wings and tail and a larger head than males. Females are also larger in size than males, but there is no overlap in size between the three species. **Of the accipiters, juvenile female Sharp-shinned Hawks and adult male Cooper's Hawks are the most similar in size and shape, and identification between the two is often difficult,** especially in western North America where Cooper's Hawks average smaller than in the East. By contract, there are obvious differences in shape between female Cooper's Hawks, which are lengthy overall, and male Sharp-shinned Hawks, which are compact. Female Cooper's Hawks are the closest in size to male goshawks, and the most likely to be confused with them.

MIGRATION

The dynamics of accipiter migration differ from those of buteo. Buteos, such as Broad-winged Hawks, typically migrate in groups and appear to help each other seek lift. Accipiters migrate singularly, but on days when accipiters are numerous they may appear to follow each other in pairs or groups. Accipiters are energetic, which makes them exciting to watch on migration. Without warning, they may attack one another in mock combat or true aggression, or dive into cover in pursuit of prey.

Of the accipiters, Sharp-shinned Hawks are the most widespread and seen in the largest numbers at North American migration sites. The peak of the fall migration for Sharp-shinned and Cooper's Hawks occurs from late September to early October in the East, but it may occur as much as two weeks earlier in the West. Some years, significant Sharp-shinned Hawk flights persist into November, when the migration of Cooper's Hawks has all but ceased. Northern Goshawks migrate primarily between mid-October and late November. Some sites, especially in the West, commonly see goshawks throughout September as well. Spring accipiter migration occurs from March to early May. Goshawks peak in late March and early April, Cooper's Hawks in early to mid-April, and

Sharp-shinned Hawks from late April to early May. Juveniles peak somewhat earlier in fall and later in spring than adults.

The fall season sees the largest flights of accipiters, with particularly high totals tallied at Cape May Point, the Goshute Mountains, and Hawk Ridge (see tables 2 and 3). In spring, only the accipiter totals for Braddock Bay and Derby Hill approach the top fall counts. Kiptopeke, Hawk Cliff and Holiday Beach in Ontario, Lake Erie Metropark, Whitefish Point, Bountiful Peak in Utah, the Manzano Mountains in New Mexico, and the Appalachian Mountains experience good accipiter flights as well. Coastal sites usually register a later peak and a higher percentage of juvenile birds than inland sites. **Accipiters, especially Sharp-shinned Hawks, are notorious for taking flight at the first sign of daylight.**

PLUMAGE

Juvenile

Juvenile accipiters are buff colored below with dark streaking. On average, Cooper's Hawks are the most lightly marked of the accipiters, whereas Sharp-shinned Hawks and goshawks often show heavy streaking. There is significant variation within all three species, and the extent of streaking on the underside of accipiters may be difficult to judge in the field. **Be aware that some Cooper's Hawks, especially in the West, are extremely heavily marked underneath, whereas some Sharp-shinned Hawks and goshawks are lightly marked.** Juvenile Sharp-shinned Hawks and goshawks are known to have barred flanks and leggings, but Cooper's Hawks can show this trait as well. The markings on the underwing coverts of Northern Goshawks are usually more prominent than on Sharp-shinned and Cooper's Hawks and may appear "checkered" overall.

Juvenile accipiters are uniform dark brown above with sparse pale spotting on the upperwing coverts. Northern Goshawks are more intricate, however, showing slate- and buff-colored tones throughout the upperside. As a result, the uppersides of juvenile goshawks may appear paler (sometimes grayish) overall than those of Sharp-shinned and Cooper's Hawks. **Goshawks typically show pale mottling along the upperwing coverts that forms an obvious narrow "bar."** Sharp-shinned and Cooper's Hawks lack this trait, but some Cooper's Hawks may show mottling or fading on the upperwing that appears similar to the upperwing bar of goshawks. Northern Goshawks also have pale auriculars that form a fairly defined facial disk, and a rufous tone to the nape and upper back. **Because of their facial disks, goshawks may appear pale headed in the field.** Most Cooper's and some Sharp-shinned Hawks have a rufous nape, but they often lack a rufous upper back. Although juvenile Sharp-shinned and Cooper's Hawks can exhibit a pale superciliary, the superciliary of juvenile goshawks is always prominent.

Adult

Adult Sharp-shinned and Cooper's Hawks are nearly identical underneath. Both are whitish below with dense rufous barring (a few Sharp-shinned Hawks are completely rufous on the chest and belly). **Since the undersides of Sharp-shinned and Cooper's Hawks are almost identical, distinguishing the two species based on plumage is often difficult.** Northern Goshawks are pale gray with fine dark barring underneath, appearing pale overall. Often, the pale underwing coverts contrast with the darker flight feathers, making the underside of adult goshawks somewhat two-toned. Some goshawks in their first year of adulthood, acquired at about one year old, show more extensive barring than older adults. The extent of barring on adult accipiters varies slightly between individuals, but even relatively heavy barring often appears faint in the field. All adult accipiters have pure white undertail coverts which contrast with the body and tail.

Adult accipiters are blue-gray above. In general, adult Northern Goshawks are slightly paler and more silvery blue than Sharp-shinned and Cooper's Hawks. Cooper's Hawks have a dark gray cap that contrasts with a paler nape; Sharp-shinned Hawks have a dark cap but lack the pale nape. The dark cap and cheeks of goshawks are bisected by a white superciliary, which often extends to the back of the head. Adult and juvenile accipiters have a banded tail, but the banding is often indistinct and difficult to observe in flight; however, the white tail tip can be a useful trait for separating accipiters. The white tail tip on Cooper's Hawks and goshawks is typically more prominent than that of Sharp-shinned Hawks. The overall plumage of accipiters, especially the upperside, fades by spring and is paler than usual. Adults with faded uppersides may appear brownish on top, similar to juveniles. In contrast, juveniles in fresh fall plumage may show a grayish sheen above similar to that of adults, especially at eye level in bright sunlight.

Although accipiters are not considered sexually dimorphic, plumage differences between adults do occur. **Adult males have pale bluish upperwing coverts with contrasting darker hands, appearing somewhat two-toned overall compared to the uniform grayish upperside of females.** This is the most useful plumage trait when separating adult male Cooper's Hawks from adult female Sharp-shinned Hawks from above. Some female accipiters have a bluish back, but their upperwings are uniformly slate colored. By contrast, some adult females can be quite brownish on top. The upperwings of some adult male goshawks can be quite pale and appear almost whitish in direct bright sunlight, especially from a head-on view. Their primaries, however, may appear blackish in poor lighting or when side-lit. Sexing adult goshawks by plumage can be difficult since some females can appear two-toned, similar to males.

From the underside, the base of the wings of adult male Sharp-shinned and Cooper's Hawks often appears silvery, in part because of the bluish topsides of the secondaries. Adult male Cooper's Hawks have gray cheeks, but many males in their first year of adulthood have rufous cheeks. Sharp-shinned Hawks and female Cooper's Hawks typically possess rufous cheeks, but a few individuals can have grayish cheeks. **The underside of adult male Cooper's Hawks can be noticeably more vibrant orange than that of adult female Cooper's Hawks and adult Sharp-shinned Hawks.**

SIMILAR SPECIES

Since juvenile Red-shouldered, Broad winged, and Red-tailed Hawks appear similar to juvenile Northern Goshawks, they are sometimes mistaken as such. Broad-winged Hawks are equally confused with Cooper's Hawks; however, they are slightly darker above with stockier, more pointed wings and a shorter, narrower tail than Cooper's Hawks and goshawks. The wing beats of Broad-winged Hawks are stiff and "choppy" compared to the more fluid wing beats of accipiters. Red-shouldered Hawks display narrow, translucent comma-shaped "windows" toward the tips of their primaries and lack the tapered wings of goshawks. The wing beats of Red-shouldered Hawks are often described as accipiter-like; however, they are shallower and less rigid along the hands than those of Northern Goshawks. Juvenile Broad-winged and Red-shouldered Hawks are typically less heavily marked underneath than juvenile goshawks.

Juvenile intermediate and dark morph Red-tailed Hawks are heavily streaked underneath and may appear similar to goshawks. Unlike on goshawks, the streaking on "dark" Red-tailed Hawks is extremely dense, making them appear overall dark underneath instead of pale overall like goshawks. All juvenile Red-tailed Hawks (and juvenile Broad-winged Hawks) have pale primary wing panels, which goshawks lack. In all postures, the wings of Northern Goshawks are stockier and the tail is longer compared with Red-tailed Hawks. Goshawks also lack the dark wing tips that buteos show. From above, juvenile Red-shouldered, Broad-winged, and Red-tailed Hawks often exhibit pale mottling along the upperwing coverts similar to juvenile goshawks, but juvenile buteos typically show pale uppertail coverts whereas **juvenile goshawks show dark uppertail coverts.**

ACCIPITER PITFALLS

Plumage

When seen well, the underside plumage of juvenile accipiters can be helpful in identification, but it should be noted with caution. Since juvenile Northern Goshawks typically have more prominent streaking on the underside than Cooper's Hawks, they are often described as "dark" underneath. Although some goshawks have broad, blobby streaking underneath and appear somewhat dark, many display narrow streaking on the underside and appear pale overall. In rare cases, lightly marked juvenile goshawks can lack markings on the undertail coverts. **Although moderately to heavily marked juvenile Northern Goshawks always have streaked undertail coverts, even extremely heavily marked Cooper's Hawks do not.** Regardless, the markings on the undertail coverts of juvenile goshawks are often difficult to observe on flying birds.

Be aware that many juvenile Sharp-shinned Hawks have dense, rufous brown streaking underneath that appears similar to the rufous barring on the underside of adults. These densely marked juveniles can be particularly difficult to age when high overhead. Juvenile and adult Sharp-shinned Hawks can also be difficult to tell apart when the sun is low on the horizon. In this setting, the wings may shadow the body while the undertail coverts remain illuminated, showing a sharp contrast similar to adults. When overhead toward sundown, the buff-colored underside of juvenile Sharp-shinned and Cooper's Hawks may appear rufous like that of adults.

Head Projection

A well-known characteristic of Sharp-shinned Hawks is their "small" head. It is indeed smaller than that of Cooper's Hawks and goshawks, but it still projects past the leading edge of the wings in most postures. **Be aware that Sharp-shinned Hawks with a full crop display an exaggerated head projection and thus appear more similar to Cooper's Hawks.**

FLIGHT STYLE

Accipiters are often direct in flight, soaring somewhat less during migration than other raptors. At coastal or flatland sites, accipiters frequently rely on powered flight, flapping and gliding intermittently, especially during high winds or cloud cover when thermals cannot form. Be careful; all raptors flap and glide intermittently on occasion, especially at ridge sites on days with light winds. When lift is sufficient, accipiters may glide for long periods without flapping.

When soaring, Sharp-shinned Hawks and Northern Goshawks hold their wings flat. On rare occasions, these two species exhibit a slight dihedral; **Cooper's Hawks almost always soar with a slight dihedral.** Sharp-shinned Hawks are extremely buoyant and rise quickly in a soar, especially on strong ridge updrafts. At low altitudes, Sharp-shinned Hawks turn in tight circles, but once up high they may circle more lazily like Cooper's Hawks and goshawks. In moderate to high winds, **Sharp-shinned Hawks appear hyperactive, unstable, and hesitant, making constant wing adjustments.** Only in very light winds, or when gliding off-ridge into a head wind, do they appear steady. Cooper's Hawks and Northern Goshawks appear steady in almost all conditions. **In very high winds along a ridge, Sharp-shinned Hawks often "dart" by in a full tuck; Cooper's Hawks and Northern Goshawks do not.**

Wing Beat

In general, smaller birds such as Sharp-shinned Hawks flap more quickly than larger birds such as Northern Goshawks; however, it is the manner in which each species flaps that is distinctive. **Sharp-shinned Hawks beat their wings in a shallow, snappy, powerless manner, similar to a robin.** Their flight style seems hurried and anxious, and when halting their wing beats, Sharp-shinned Hawks appear hesitant to do so; Cooper's Hawks and goshawks lack hesitation. Even when flapping aggressively while migrating, Sharp-shinned Hawks often move relatively slowly. However, they can move at high speeds when chasing prey. **The wing beats of Cooper's Hawks appear as shallow as those of Sharp-shinned Hawks but are stiffer and more forceful.** Cooper's Hawks appear deliberate in flight compared with the anxious manner of Sharp-shinned Hawks. Northern Goshawks' wing beats are **loftier (deeper on the upstroke)** and more effortless than those of Cooper's Hawks. **When flapping, the wings of goshawks appear straighter, with less motion at the wrists than in other accipiters.** Female goshawks can exhibit loose, heavy wing beats similar to those of buteos.

The rate at which accipiters beat their wings can vary on occasion. For example, Sharp-shinned Hawks can flap relatively lazily while soaring in light winds, whereas Northern Goshawks, especially males, can flap quickly when lift is minimal. Male Cooper's Hawks can appear to flap as quickly as female Sharp-shinned Hawks under any conditions. When flapping into a strong head wind, accipiters often draw their hands in and appear pointed winged, similar to falcons. With the exception of American Kestrels, however, falcons are extremely fast moving when flapping vigorously. American Kestrels and Sharp-shinned Hawks have very similar flight styles, but during powered flight **the hands of American Kestrels sweep back in the rhythmic fashion of a sea turtle, whereas Sharp-shinned Hawks flap in a more up-and-down fashion.**

Sharp-shinned Hawk

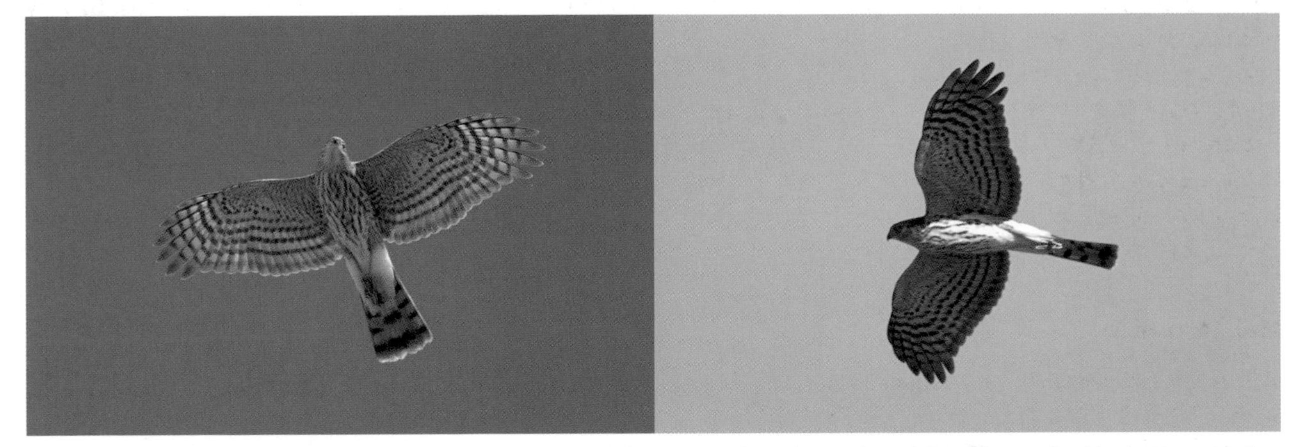

SS 01 - Sharp-shinned Hawk, juvenile. Typical plumage (left, UT) is pale underneath with heavily streaked body. Some birds (right, NV) can be more lightly marked than Cooper's Hawks; however, they are males, which are very different in shape and flight style than Cooper's Hawks. Note stocky wings, small head, and short, narrow tail.

SS 02 - Sharp-shinned Hawk. Juveniles (left, NJ) that are heavily streaked underneath can appear barred, similar to adults (right, NV). Note evenly spaced, pale rufous barring on underside of adult.

SS 03 - Sharp-shinned Hawk, adult (NV). A few adults are solidly rufous on the body, which Cooper's Hawks do not show.

SS 04 - Sharp-shinned Hawk, juvenile (NV). Brown above with gray and black tail bands. Note small head, narrow base to tail, and white tail tip.

SS 05 - Sharp-shinned Hawk, adult. Adult males (left, NV) are blue above with contrasting darker hands. Adult females (right, NV) are uniformly gray-blue on top. Note broad, white tail tip of male, similar to tail tip of Cooper's Hawks.

SS 06 - Sharp-shinned Hawk. Adults in spring (left, MI) are often faded and appear brownish, similar to juveniles (right, NJ). Note contrast between brownish hands and bluish back of adult; juveniles always show uniformly brown upperside. Photo on left © Michael Shupe

SS 07 - Sharp-shinned Hawk. In spring, when birds are paler than usual, rufous streaking of juveniles (left, MI) can appear similar to rufous barring of adults (right, MI). Note adult's shorter tail and straighter trailing edge on wings.

SS 08 - Sharp-shinned Hawk, juvenile (UT). Sharp-shinneds with full crop show a greater head projection than usual. Note stocky wings and narrow tail.

SS 09 - Sharp-shinned Hawk, adult (NV). In strong winds along a ridge, Sharp-shinned Hawks often pass in a high-speed glide with wings pulled in.

Cooper's Hawk

CH 01 - Cooper's Hawk, juvenile (NJ). Typical plumage (left) with moderately streaked body and tawny head. Lightly marked birds (right) show streaking limited to chest. Note long tail with rounded tip.

CH 02 - Cooper's Hawk, adult (NV). Adult Cooper's Hawks show orange and white barring underneath with white undertail coverts. Males (left) are often vibrant orange and show grayish cheeks and nape. Adult females (right) are less vibrant underneath and typically show rufous cheeks.

CH 03 - Cooper's Hawk. Juveniles (left, UT) are brownish on top with tawny head. Note long tail with broad, white, rounded tip. Some adult females (right, NV) can appear brownish on top, similar to juveniles. Note dark cap, rufous nape, and flight-feather molt. In spring, this adult would be extremely brownish because of fading.

CH 04 - Cooper's Hawk, adult (NV). Adult males (left) are blue above with contrasting darker hands and grayish cheeks and nape. Adult females (right) are unformly gray-blue on top with rufous cheeks and nape. Note long tails with rounded tips.

CH Pitfall 01 (NV). Compare uniform upperside of adult female **Sharp-shinned Hawk** (left) with contrasting upperside (bluish upperwings and blackish hands) of adult male **Cooper's Hawk** (right). Note larger head and longer, broader, rounded-tipped tail of Cooper's Hawk. Shown same size for plumage comparison.

CH Pitfall 02. Cooper's Hawks (left, UT), such as this adult male (vibrant underside, gray cheek, dark cap), can show a square-tipped tail because of molt. By contrast, **Sharp-shinned Hawks** (right, NV) can have a rounded tail tip; also note stocky wings and body and rufous streaking on chest.

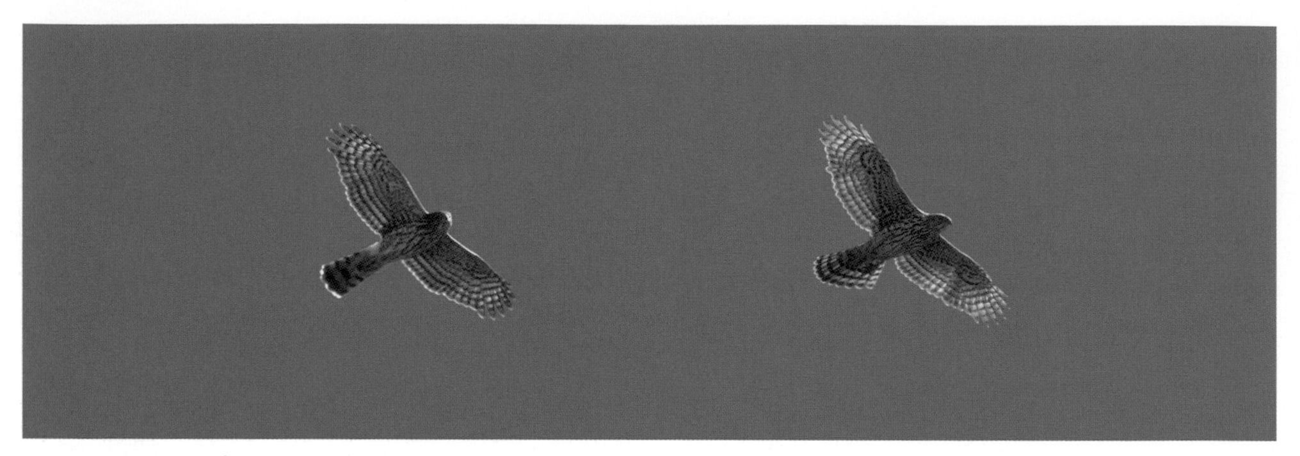

CH Pitfall 03. At a distance or when backlit, heavily marked juvenile **Cooper's Hawks** (left, NV) can be difficult to tell from juvenile **Northern Goshawks** (right, MI) based on plumage. Note broad wings and body of goshawk. Birds shown same size for plumage comparison.

Northern Goshawk

NG 01 - Northern Goshawk, juvenile. Lightly marked juvenile goshawks (left, NV) can resemble Cooper's Hawks. Typical goshawks (right, MI) are heavily marked underneath. Note wedge-shaped tail of bird on left and broad wings and body of both birds.

NG 02 - Northern Goshawk, adult (NV). Note pale grayish underside with contrasting darker flight feathers and white undertail coverts.

NG 03 - Northern Goshawk, juvenile (NV). Brownish above with pale "upperwing bar," rufous nape, and pale eye-line.

NG 04 - Northern Goshawk, juvenile (NV). Some juveniles lack pale "upperwing bar" and obvious rufous nape. Note broad wings and back.

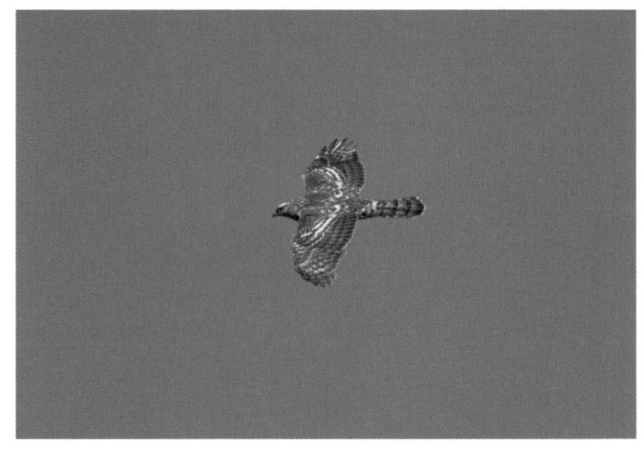

NG 05 - Northern Goshawk, juvenile (UT). Juveniles can show distinctly whitish "upperwing bar" in spring because of fading.

NG 06 - Northern Goshawk, adult. Adult males (left, NV) are pale blue on top with blackish hands. Females (right, MN) are uniformly dark gray-blue above and show less pointed wings than males. All adults have blackish head with distinct white eye-line. Photo on right © Brian K. Wheeler

NG Pitfall 01. Adult female **Cooper's Hawks** (left, UT) and adult male **goshawks** (right, NV) are both grayish on top and similar in size; however, adult male goshawks are two-toned above compared with uniform tone of adult female Cooper's Hawks. Birds shown same size for plumage comparison.

NG Pitfall 02. Adult female goshawks (NJ) can be quite brownish on top when they retain a significant amount of faded adult feathers.

SOARING

While soaring, accipiters exhibit stocky wings and a long tail compared with other raptors. Sharp-shinned Hawks have short, stocky, squared-off wings that often taper at the body and bow forward at the leading edge. The wings of Cooper's Hawks are longer, slimmer, and do not bow forward. **The wings of Northern Goshawks are broad at the base with relatively long, tapered hands; this creates a more angular trailing edge to the wings than on Sharp-shinned and Cooper's Hawks.** Juvenile goshawks have a broader wing base than adults, creating a somewhat more angular appearance to their wings. Sharp-shinned Hawks have a small, narrow head and stocky body. Cooper's Hawks and goshawks have a larger head that protrudes farther past the leading edge of the wings than in Sharp-shinned Hawks. The body of goshawks usually appears broad overall, whereas the body of Sharp-shinned Hawks appears short and stocky at the chest. Cooper's Hawks have a relatively slim body.

Although the tail tip of all accipiters is rounded when spread, accipiters often soar with their tail folded. When folded, Sharp-shinned Hawks often exhibit a square-tipped tail, but females can have a rounded tail tip. Cooper's Hawks typically have a rounded tail tip and Northern Goshawks a rounded or wedge-shaped tail tip. **Regardless, the tail of Sharp-shinned Hawks is shorter, and narrower at the base, than that of Cooper's Hawks and goshawks.**

Most adult Sharp-shinned and Cooper's Hawks in eastern North America complete their molt before they migrate south in fall. Therefore, adult accipiters with a square-tipped tail are almost always Sharp-shinned Hawks. In contrast, many Western accipiters are still actively molting during fall, and it is common to see adult Cooper's Hawks with a square-tipped tail. **Since juvenile accipiters do not molt until their first spring, juveniles with a square-tipped tail in fall are almost always Sharp-shinned Hawks. Be aware that in spring, most accipiters have a square-tipped tail as a result of feather wear.** Be careful when using the width of the white terminal band on the tail to distinguish accipiters. Variation in width occurs naturally, and when backlit the white tail tip often appears unusually bright or broad.

HEAD-ON

Head-on identification of raptors can be tricky, and thus separating the accipiters can be particularly difficult. From this angle, Sharp-shinned Hawks and goshawks appear deep chested, whereas Cooper's Hawks display a wide, shallow body. From slightly above, the back of Sharp-shinned Hawks appears narrow, whereas Cooper's Hawks and goshawks exhibit a broad, powerful back.

All accipiters glide with a slight droop to their wings. **Since goshawks have the longest hands of the accipiters, they have the most pronounced droop to their wings.** Of the accipiters, Sharp-shinned Hawks possess the stockiest wings and appear least droopy. Also, the wing tips of Sharp-shinned Hawks are the most squared off of the accipiters, those of goshawks the least. This makes goshawks, especially adult males, appear extremely falcon-like when viewed head-on. **Adult male Sharp-shinned and Cooper's Hawks may look pointed winged as well, which is often a giveaway when telling adult male Cooper's Hawks from female Sharp-shinned Hawks.** Sharp-shinned Hawks that appear pointed winged from this angle resemble American Kestrels. However, kestrels display a slimmer chest and longer, slimmer, more drooped wings than Sharp-shinned Hawks. On windy days along a ridge, kestrels tend to flap more often than Sharp-shinned Hawks.

GLIDING OVERHEAD

All accipiters display stocky wings and a long tail when gliding overhead. Primary projection is helpful in distinguishing among the three accipiters from this angle. **Goshawks have the longest hands of the accipiters and thus have the greatest primary projection in a glide,** whereas Sharp-shinned Hawks have the least. The wings of goshawks are also the broadest of the three accipiters; however, the relatively slim wings of adult male goshawks make them appear the most falcon-like of the accipiters in a glide.

Sharp-shinned Hawks have a stockier body and wings than Cooper's Hawks, but the tail of Cooper's Hawks is longer and broader at the base; Sharp-shinned Hawks always show a narrow tail. The head of Sharp-shinned Hawks appears rounded and narrow, almost turtlelike; of Cooper's Hawks and goshawks is triangular and stocky, like that of buteos. Observing the overall silhouette of overhead accipiters is the most accurate way to separate them.

WING-ON/GOING AWAY

From a side view, the wings of Sharp-shinned Hawks appear short, broad, and somewhat squared off. The head projects slightly past the leading edge of the wings, the body is stocky, or "chesty," and the back is narrow. Goshawks are similar in shape to Sharp-shinned Hawks, but their wings are broader at the base and more tapered at the hands, creating an exaggerated bulge along the trailing edge. Goshawks show a broad, powerful back and relatively large head. The trailing edge on the wings of Cooper's Hawks is straighter than in Sharp-shinned Hawks and goshawks. This is most apparent on birds headed away. In addition, Cooper's Hawks exhibit a slimmer body than Sharp-shinned Hawks and Northern Goshawks. Wing-on, the head of Cooper's Hawks projects well past the leading edge of the wings, giving the wings and head combined a more triangular silhouette than that of Sharp-shinned Hawks.

When viewing the topside of adults, remember that gender-related plumage traits can be extremely helpful in telling female Sharp-shinned Hawks from male Cooper's Hawks. Observing plumage traits of juveniles, especially the width of the white tail tip, may also be helpful from above. **The tail tip of all accipiters appears squared off when headed away; however, the tail of Sharp-shinned Hawks looks especially shortened.** Also, when gliding away, all raptors exhibit pointed wing tips.

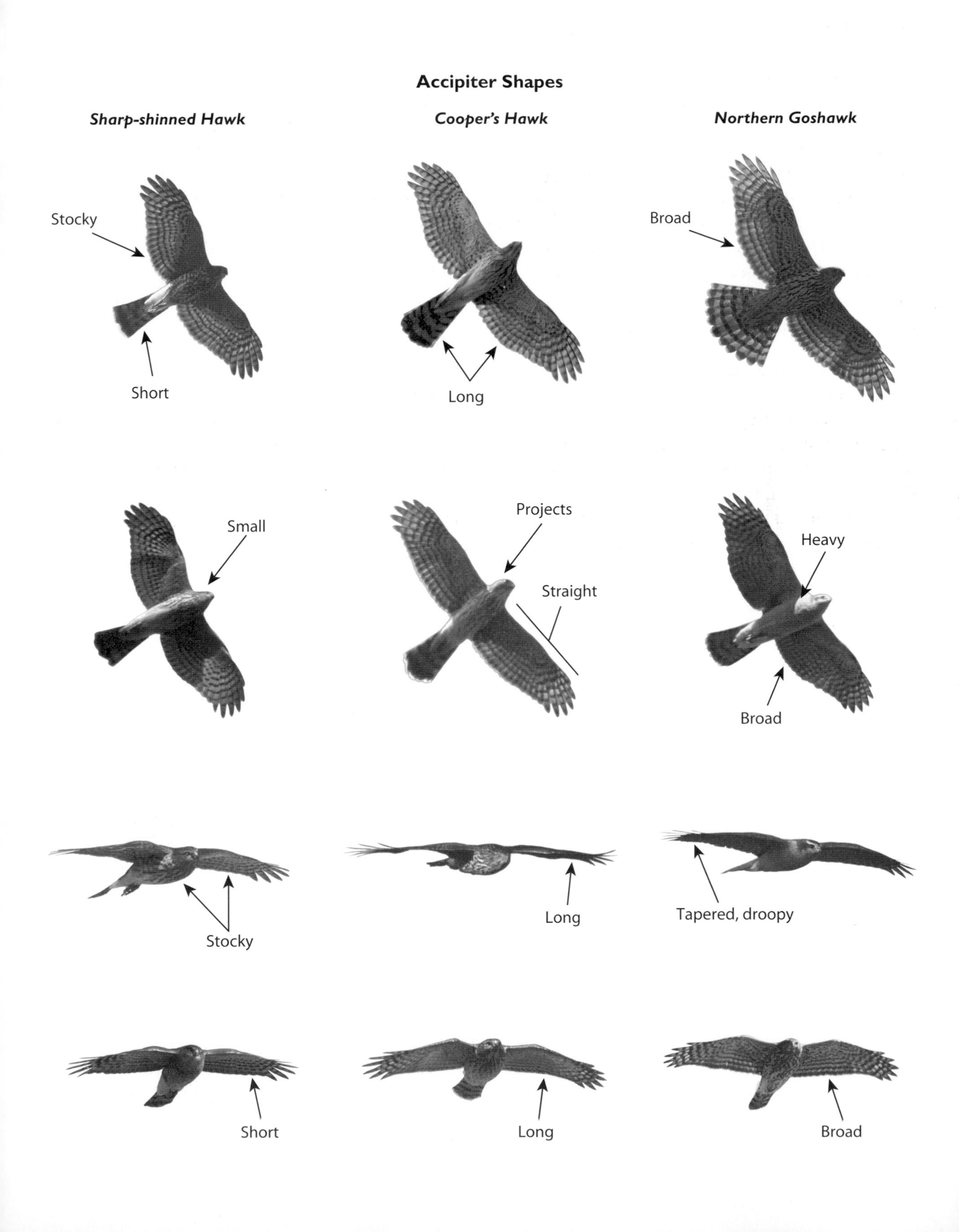

Accipiter Shapes
Sharp-shinned Hawk
Cooper's Hawk
Northern Goshawk
Stocky
Short
Long
Broad
Small
Projects
Straight
Heavy
Broad
Stocky
Long
Tapered, droopy
Short
Long
Broad

Accipiter Shapes

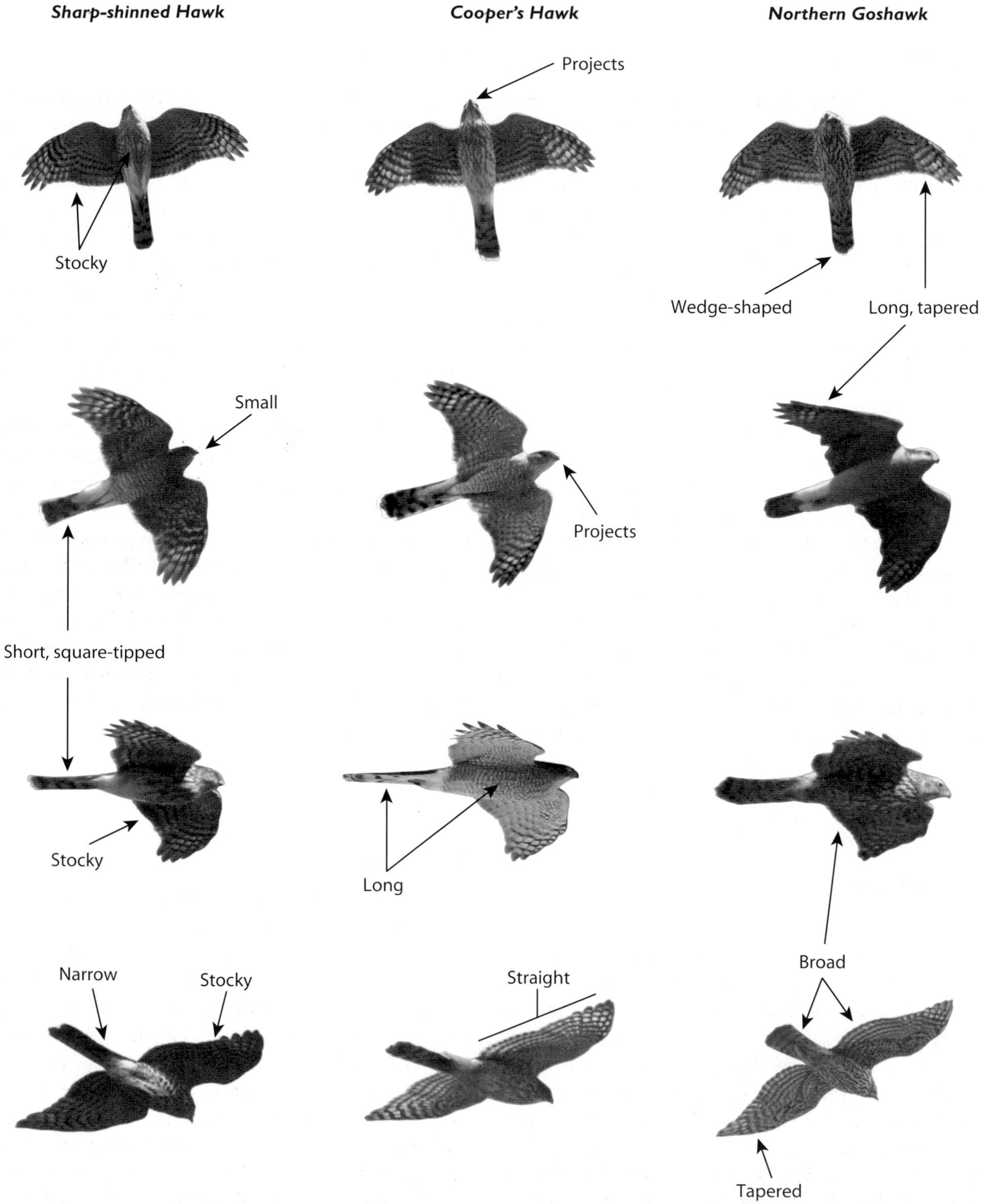

Sharp-shinned Hawk | *Cooper's Hawk* | *Northern Goshawk*

Projects

Stocky

Wedge-shaped

Long, tapered

Small

Projects

Short, square-tipped

Stocky

Long

Broad

Narrow

Stocky

Straight

Tapered

Northern Harrier (*Circus cyaneus*)

OVERVIEW

The only North American representative of the worldwide genus *Circus*, the Northern Harrier occurs throughout eastern and western North America. Often compared to Short-eared Owls, with which they share the same marshes and grasslands, harriers are capable of locating prey by sound in near darkness. This is due in part to their acute hearing, which is enhanced by a ring of stiff feathers surrounding their face, called a facial disk, which deflects sound toward their ears. Although they can hunt in darkness, harriers are most active during the daylight hours, when they seek prey using both sight and sound.

The most well-known and noticeable field mark of Northern Harriers is the patch of brilliant white uppertail coverts, commonly referred to as the "white rump." This marking can be very helpful in identifying low-flying harriers, as it can be seen at great distances. However, it can also be confused with similar field marks of other raptors (see Harrier Pitfalls).

Northern Harriers are distinctive in flight and can often be identified by their flight style alone. They fly much like Turkey Vultures, constantly teetering from side to side, but are smaller and slimmer overall. Despite their wobbly demeanor, Northern Harriers are agile fliers and can out-maneuver other large raptors when threatened. Harriers can vary in shape radically, appearing falcon-like in a glide or buteo-like in a soar.

Size and Structure

Northern Harriers have long, narrow wings and tail, exhibiting an overall lanky appearance. Although large in frame, they are extremely lightweight and buoyant. Males are smaller than females, but this difference in size is only noticeable in the field when both sexes are seen in close proximity to each other. Males have proportionately shorter wings and tail than females, which is noticeable in flight only with considerable practice. Also, males generally hold their wings in a slightly shallower dihedral than females, perhaps because of their shorter wingspan.

MIGRATION

Northern Harrier migration occurs throughout the entire spring and fall seasons. Harriers can be seen moving south from August through December in fall. In spring they head north as early as February and as late as June. There is a fair distinction between the peak time periods for juveniles and adults. In fall, the peak for juvenile birds usually occurs from mid- to late September, with the adult peak often following several weeks later. In spring, the peak time for harriers is early to mid-April; juveniles migrate throughout the spring, with a gradual peak from mid-April to early May.

Northern Harriers can be seen migrating from the first sign of light until darkness; they have even been seen migrating at night! At Cape May, I once saw a group of 16 harriers crossing Delaware Bay at sunrise. Braddock Bay holds the daily high count of 440 and seasonal high count of 3,177 for North America, but Cape May Point averages the most harriers per season, with 1,750. Derby Hill, Whitefish Point, Hawk Ridge, Lake Erie Metropark, Holiday Beach, Kiptopeke, Bountiful Peak, Hawk Mountain, and the Niagara Peninsula in Ontario are all excellent sites to observe harriers on migration.

PLUMAGE

Adult Female and Juvenile

Adult female and juvenile Northern Harriers are similar in plumage. Juveniles have a rufous underbody with faint dark streaking on the chest, which is typically not visible in flight. Adult females are buff underneath with dark streaking throughout the body; however, **many adult females in western North America have a rufous tone on the underbody similar to that of juveniles**. The axillaries, underwing linings, and secondaries of adult females and juveniles are dark and contrast with the slightly paler primaries. **In spring, the undersides of most juveniles fade to buff, causing them to look almost identical to adult fe-**

males. Even at close distances, separating adult females from juveniles can be challenging.

Juvenile and adult female Northern Harriers are brown above with sparse pale mottling along the upperwing coverts. The mottling on juveniles is rufous but often fades to buff by spring. Adult females show either buff mottling on the upperwing or rufous mottling that fades to buff by spring. The topside of the remiges of adult females is brown with grayish tones that may contrast slightly with the dark brown upperwing coverts. **Juveniles have dark brown flight feathers and appear even toned throughout the upperside.** The head of juvenile harriers is dark brown with a pale, incomplete eye-ring. On adult females, the head is slightly paler brown with more extensive mottling throughout. This is most noticeable when viewed at eye level. The tail of adult female and juvenile harriers is dark brown with several faint pale bands. **All Northern Harriers have sharply contrasting bright white uppertail coverts that are visible at great distances.**

Adult Male

Adult male Northern Harriers are unique among North American raptors. **They are bright white underneath with black wing tips and a black trailing edge to the secondaries.** Adult males have a gray "bib" that, when combined with the gray head, forms a complete hood. Some males have rufous barring on the chest, but the barring is often faint in the field. The upperside of adult male harriers is pale gray with faint, pale mottling on the upperwing coverts.

The coloration on the upperside of adult male harriers varies slightly between individuals. Some males are brownish gray, especially in their first year of adulthood, whereas others are slate gray or pale gray. Birds in spring are often paler than usual on top because of fading. When viewed head-on, adult males that are particularly slate colored on top may appear brownish, similar to adult females or juveniles.

HARRIER PITFALLS

"White Rump"

The white uppertail coverts, or "rump patch," on Northern Harriers can be confused with similar field marks of other birds. Light morph Swainson's Hawks and most light morph Red-tailed Hawks have pale uppertail coverts than can appear similar to those of harriers. However, **the "rump patch" of Northern Harriers is gleaming white, broad, and cut squarely across the base of the tail.** The pale uppertail coverts on Swainson's and Red-tailed Hawks are less brilliant and form a narrow U shape across the base of the tail.

Birds with a white tail base, such as light morph Rough-legged and Ferruginous Hawks and immature Golden Eagles, are sometimes mistaken for Northern Harriers. The white tail base of buteos and eagles is more prominent, and wider than the "rump patch" of harriers. Also, there are differences in overall plumage, shape, and flight style between buteos, eagles, and harriers. **Caution: Be careful when observing birds from a side view, as the white undertail coverts of many raptors can be confused for the white "rump patch" of Northern Harriers.** Be aware that the white uppertail coverts on Northern Harriers are not visible from below.

Dark Flight Feathers

When seen high overhead, the dark axillaries, underwing linings, and secondaries of adult female and juvenile Northern Harriers can be confused with the dark axillaries and underwing linings of Prairie Falcons. Although harriers can look dark throughout the entire basal half of the wings, the dark area on the underside of Prairie Falcons is limited to the axillaries and underwing linings. Prairie Falcons are pale otherwise and do not exhibit dark hoods like harriers. Because female and juvenile harriers have dark secondaries, they can appear similar to Swainson's Hawks, but Swainson's Hawks are dark throughout their flight feathers.

FLIGHT STYLE

Northern Harriers are extremely buoyant and agile in flight. They can gain lift on even the slightest breeze, often appearing to fly leisurely in conditions where other raptors seem to struggle to stay aloft. During conditions that keep raptors at relatively low altitudes, harriers are often the highest-flying hawks in the sky. Harriers that are high overhead can fool beginning birders who are accustomed to seeing them skimming the grasslands and marshes while hunting. Harriers soar lazily like large buteos but are direct when migrating. It is rare to see them circle back at peninsular sites or stop to rest elsewhere. With relative ease, harriers readily use powered flight to cross large bodies of water or desert expanses. Northern Harriers rarely flap continuously for long periods of time, even during conditions of poor lift that cause other raptors to do so.

Many raptors soar with a dihedral, but **Northern Harriers exhibit a more pronounced dihedral than most;** only Turkey Vultures match the dihedral shown by harriers. **Harriers even teeter from side to side in flight like Turkey Vultures,** but they do so in a slightly fluid, more controlled manner. When faced with strong winds along a ridge top that are ideal lift for most species, harriers often choose to head off-ridge into the wind or perpendicular to it. When doing so, they often fly in an extremely steady manner on drooped wings. Northern Harriers are able to make tight, quick turns, but they generally move quite slowly in flight.

Wing Beat

Northern Harriers display **lofty, easy wing beats.** They are distinctly loose and deep, lacking the stiffness exhibited by most other raptors. At times, Swainson's (especially juveniles) and Ferruginous Hawks flap with deep wing beats, yet their wing beats appear stiffer and less even in tempo than those of harriers. Although some raptors, such as Northern Goshawks and Red-tailed Hawks, flap slowly or quickly under certain circumstances, the wing beats of Northern Harriers are always slow.

Pitfalls

Cooper's Hawks and Northern Goshawks normally display deliberate wing beats, but during courtship they display deep, lazy wing beats similar to those of Northern Harriers. The wing beats of Cooper's Hawks and Northern Goshawks during courtship are mechanically stiff, with the wings appearing to touch at the top of each upstroke. It is not uncommon to see Cooper's Hawks performing courtship displays during spring migration. Be aware that during courtship, Cooper's Hawks and goshawks also spread their undertail coverts, which can appear like a white "rump patch."

Northern Harrier

NH 01 - Northern Harrier, juvenile (NJ). Juvenile harriers are rufous underneath in fall (left), fading to buff by spring (right). The faint streaking on chest of juveniles is typically inconspicuous in the field.

NH 02 - Northern Harrier, adult female. Adult females are pale underneath with dark streaking (left, WA); however, some adults (right, UT), especially western birds, are particularly rufous underneath, similar to juveniles. Although heavily streaked, the bird on right can appear solidly rufous at a distance.

NH 03 - Northern Harrier. In spring, juveniles (left, NJ) and adult females (right, MI) can look extremely similar because of fading. At a distance, they can be impossible to tell apart when streaking on adults is indistinct.

NH 04 - Northern Harrier, adult male (WA). From below (left), adult male harriers are white with gray hood and show dark wing tips and dark terminal band along secondaries. From above (right), adult males are grayish overall. All harriers display brilliant white "rump." Photo on left © Sherry Liguori

NH 05 - Northern Harrier, juvenile. In fall, topside of juvenile harriers (left, UT) is uniformly dark brown with rufous mottling along upperwing coverts. Upperwing of juveniles fades by spring (right, MI), appearing similar to that of adult females but still uniform in tone.

NH 06 - Northern Harrier, adult female. Typical adult females (left, NV) are brown above with grayish and buff tones throughout. Some adult females (right, same bird as NH 02), especially birds with rufous undersides, are particularly dark on top, similar to juveniles, but still show a grayish cast to remiges. Photo on left © Chris Neri

NH Pitfall 01 (UT). White-based tail of **Ferruginous Hawks** (left) and **Rough-legged Hawks** (right) can be mistaken for white "rump" of harriers but is broader and more extensive. Pale flight feathers of most buteos and white primary wing panels of Ferruginous Hawks are not shown by harriers.

NH Pitfall 02. **Swainson's Hawks** (left, CO) and **Red-tailed Hawks** (right, UT) often exhibit pale uppertail coverts that resemble white "rump" of harriers; however, they are narrower and less brilliant. Note two-toned upperside and pointy wings of Swainson's Hawk and broad wings of Red-tailed Hawk. Photo on right © Sherry Liguori

NH Pitfall 03 (NV). Immature **Golden Eagles** have white-based tail similar to white "rump" of harriers; however, Golden Eagles show broader wings overall.

NH Pitfall 04 (UT). From wing-on, white undertail coverts of raptors such as this adult **Cooper's Hawk** can be mistaken for uppertail coverts of harriers.

NH Pitfall 05 (UT). From eye level, **harriers** (left) and **Rough-legged Hawks** (right) may appear identical in shape; however, notice stocky body and slightly shorter hands of Rough-legged Hawks. Most Rough-legged Hawks show pale chest and pale leading edge to wings. Photo on right © Chris Neri

SOARING

When soaring, the long, slim wings and tail of Northern Harriers give them one of the most distinctive silhouettes of all raptors. The wings of harriers do not show a bulge along the trailing edge like those of many other raptors, and the head appears somewhat small. Many buteos appear long winged like Northern Harriers, but the wings of buteos are always broader than those of harriers. **While soaring, Northern Harriers always hold their wings in a strong dihedral.**

Swainson's Hawks can be particularly slim winged for a buteo and sometimes sway from side to side like harriers. However, Swainson's Hawks have sharply pointed wing tips and a shorter tail than harriers. While soaring, juvenile Red-tailed Hawks may resemble Northern Harriers, but Red-tailed Hawks are stockier overall, show a less pronounced dihedral, and are always steady in flight. Ferruginous Hawks display long wings, but their wings are broader at the base, appearing more angular than those of harriers. Unlike Northern Harriers, Ferruginous Hawks have a very broad chest.

Rough-legged Hawks are the most similar in shape to Northern Harriers. Male Rough-legged Hawks are relatively slim winged, but their wings are slightly broader overall than those of harriers. Rough-legged Hawks also possess a shorter tail and stockier body than harriers. Rough-legged Hawks may display a modified dihedral when soaring, but they typically hold their wings in a shallow dihedral.

Accipiters have a long, narrow tail, but their wings are stocky in comparison to harriers, and they never fly with a strong dihedral. Falcons possess long, narrow wings, but their wings are always pointed; harriers have slim wings that taper slightly at the hands, but they are never pointed in a soar. Despite having some similarities in flight style with accipiters and buteos, Northern Harriers are lightweight and able to gain lift more easily when soaring.

HEAD-ON

Identifying Northern Harriers approaching at eye level can be tricky. At shorelines, certain conditions, such as strong winds that prevent thermals and push migrants toward water barriers, contribute to low-altitude flights. Under these conditions, harriers fly in characteristic fashion, tipping from side to side as they battle the wind, particularly when flying perpendicular to it. On light to moderate winds along a ridge, they fly with wings held in a modified dihedral and exhibit the familiar "tipsy" flight style. **On strong ridge updrafts, harriers often fly on droopy wings.**

Harriers can be deceiving in this posture, since they typically fly with a strong dihedral. Moreover, harriers often fly more steadily than usual when gliding along an updraft.

When viewed head-on, Northern Harriers appear similar to several buteos, including Swainson's, Ferruginous, and Rough-legged Hawks. The wings of Swainson's Hawks show a gentler bow and are less sharply raised at the shoulders than those of harriers. Also, the wing tips of Swainson's Hawks are sharply pointed and held level with or slightly below the body when gliding. Northern Harriers show a slim body compared with all buteos. Ferruginous Hawks are extremely buoyant for their size and often teeter from side to side like Northern Harriers; they have a much broader chest than harriers, however. Ferruginous Hawks also hold their wings in a modified dihedral with wing tips raised above the body when gliding.

Rough-legged Hawks are the most similar in shape to Northern Harriers. However, the wings of harriers are slightly shorter at the base and longer at the hands than in Rough-legged Hawks. Rough-legged Hawks tend to hold their wing tips above the body at all times, whereas harriers will droop their wings on strong ridge updrafts. **The most significant difference between Rough-legged Hawks and harriers is the broader chest of Rough-legged Hawks.** Light morph Ferruginous, Rough-legged, and some Red-tailed Hawks show a pale head and a pale leading edge to the wings when viewed head-on. Of the harriers, adult males show these traits most often.

GLIDING OVERHEAD

When gliding, Northern Harriers resemble large falcons such as Peregrine and Prairie Falcons. The hands of harriers taper sharply and project well past the base of the wings, but their wing tips are not sharply pointed like those of falcons. Harriers also lack the broad head and chest shown by large falcons. **The tail tip of harriers is squared or notched; the tail of Peregrine and Prairie Falcons tapers toward the tip.** Peregrine and Prairie Falcons tend to glide at relatively high speeds; harriers typically glide at the same pace as accipiters and buteos. Northern Harriers may also resemble buteos such as Swainson's, Ferruginous, and Rough-legged Hawks when gliding but have slimmer wings and body and a longer tail.

WING-ON/GOING AWAY

Although accipiters and buteos appear stockier than usual from a side view, the lanky shape of Northern Harriers remains obvious from this angle. Buteos that can appear similar to harriers, such as Swainson's, Ferruginous, and Rough-legged Hawks, exhibit broader wings and a shorter, broader tail than harriers. When headed away, the wings of Northern Harriers appear extremely long and narrow and do not taper toward the body like those of similar buteos. Moreover, the tail and body of Northern Harriers appear somewhat shortened, but they are still narrower than those of similar buteos.

Northern Harrier Shapes

Buteos

Red-shouldered Hawk, Broad-winged Hawk, Swainson's Hawk, Red-tailed Hawk, Ferruginous Hawk, Rough-legged Hawk

OVERVIEW

Of the raptors, the buteos vary most in appearance. Most are pale underneath and dark above. However, all of the species considered herein, except Red-shouldered Hawk, exhibit dark morph plumages with a **continuum of plumage variations between light and dark.** Some buteos display distinct traits, such as the "red" tail of adult Red-tailed Hawks, which can make them easy to identify. Others, such as juvenile Broad-winged Hawks, are nondescript. In short, buteos present many guises that can make identification difficult at times.

Juvenile Broad-winged, Red-tailed, and Rough-legged Hawks have pale primary wing panels, or "windows," that appear translucent from below and help distinguish them from adults. In contrast, Red-shouldered and Ferruginous Hawks show translucent windows at all ages. Telling age, morph, or race of buteos is more difficult at certain angles where plumage details are difficult to see, such as head-on, wing-on, or going away.

Broad-winged, Red-shouldered, Red-tailed, and Rough-legged Hawks are seen throughout eastern and western North America; however, Broad-winged Hawks are much less common in the West. Red-shouldered Hawks are extremely rare west of Michigan, except on the West Coast where the California race occurs in California and Oregon. Swainson's and Ferruginous Hawks are primarily western species, but small populations of Swainson's Hawks nest as far east as Illinois. Swainson's Hawk reaches east to the Atlantic Coast during migration, with many eastern hawkwatching sites recording several each year.

Size and Structure

Although all raptors soar with proficiency, buteos are designed for soaring. With long, broad wings and a broad tail, buteos can soar for extended periods of time with great ease. Forest-dwelling buteos such as Broad-winged and Red-shouldered Hawks have shorter wings than open-country birds such as Swainson's and Ferruginous Hawks. Buteos vary in shape, resembling other buteo species or, in some instances, harriers, accipiters, falcons, or eagles.

Excluding Broad-winged Hawks, buteos are large raptors. Swainson's and Ferruginous Hawks often appear somewhat slim winged compared with other buteos, whereas Broad-winged Hawks are chunky. Male Swainson's, Ferruginous, and Rough-legged Hawks have slimmer wings than females, but sexing these species by shape alone is difficult. Juvenile Swainson's and Red-tailed Hawks possess shorter secondaries than their adult counterparts, making their wings slimmer and thus appearing lankier overall. With practice, aging Swainson's and Red-tailed Hawks is possible based on shape alone. Red-tailed Hawks are the only buteos in which juveniles show a noticeably longer tail than adults.

MIGRATION

Buteo migration is more dynamic and weather dependent than the migration of other raptors. Buteos typically do not migrate during days of light drizzle or poor weather, but they can be seen in large groups, called "kettles," during ideal conditions. They tend to seek lift via thermals or updrafts and typically avoid crossing large bodies of water or large expanses of desert that would require powered flight to do so. Of the buteos, Rough-legged Hawks are the least hesitant to use powered flight for extended periods.

Broad-winged Hawks gather in larger concentrations than any other raptor in the world. The majority of Broad-winged Hawks migrate inland instead of along the Pacific and Atlantic Coasts. They concentrate in the largest numbers at Veracruz, Mexico, where as many as 775,760 have been tallied in a single day! The fall Broad-winged Hawk migration in eastern North America has a faithful following that is a phenomenon in its own right. Many birders eagerly follow the weather during mid-September in fall or late April in spring in hopes of catching the "big day" at their local site. Swainson's Hawks occur in massive numbers at Veracruz as well; significant concentrations of Swainson's Hawks are rare to the north, however. The peak of the Swainson's Hawk migration occurs about a week later in fall and a week earlier in spring than the peak for Broad-winged Hawks.

Broad-winged and Swainson's Hawks head south earlier in fall and arrive north later in spring than other buteos. Swainson's Hawks are known to congregate in agricultural fields during early spring after arriving from Central and South America, and during late summer before heading south in fall. The Borrego Valley in California and Estancia Valley in New Mexico are reliable sites to see Swainson's Hawks in early to mid-spring, whereas the Butte Valley in California and Curlew Valley in Utah are reliable sites in late summer to watch large groups of Swainson's Hawks.

Red-shouldered Hawks are generally short- to medium-distance migrants, but some Eastern birds travel as far south as central Mexico in fall. Red-shouldered Hawks exhibit a distinct peak time period during spring and fall migration. In fall, late October to mid-November is the best time to see Red-shouldered Hawks, whereas late March to early April is the peak time to see spring migrants. Braddock Bay and Derby Hill are the most reliable sites in spring for observing good flights of Red-shouldered Hawks. In fall, sites along the Appalachian Mountains of northeastern Pennsylvania, such as Hawk Mountain, and northwestern New Jersey, such as Raccoon Ridge, along with Cape May Point, are the best places for viewing Red-shouldered Hawks.

Red-tailed Hawk migration in North America may occur in almost any month of the year. However, the bulk of the fall migration occurs from mid-October to mid-November in the East and about two weeks earlier in the West. Spring migration for Red-tailed Hawks peaks between mid-March and early April. In the East, a significant peak of adult Red-tailed Hawks occurs during mid- to late April. Many of these birds are headed toward their eastern Canadian breeding grounds. Braddock Bay, Derby Hill, Whitefish Point, the West Skyline Hawk Watch near Duluth, Minnesota, Dinosaur Ridge in Colorado, and the Wasatch Mountains in Utah are the best spring sites to witness Red-tailed Hawks. Cape May Point, Hawk Mountain, Hawk Ridge, Lake Erie Metropark, Holiday Beach, the Goshute Mountains, the Golden Gate Hawk Watch near San Francisco, Bountiful Peak, and the Manzano Mountains record excellent flights of Red-tailed Hawks each fall.

Red-tailed Hawks occur along the south shore of Lake Ontario from mid-August to early September. These flights comprise Eastern juveniles, which often disperse north after fledging. Counts comprising postbreeding-dispersal birds are conducted at Braddock Bay, where daily totals of more than 1,000 Red-tailed Hawks have been recorded. During this period, a considerable southbound migration of juvenile Red-tailed Hawks is observed in the West. The peak migration for juvenile Red-tailed Hawks occurs somewhat earlier in fall and later in spring than that of adults. Adults from northern latitudes and high altitudes are more likely to migrate than other adults.

Ferruginous Hawks are short- to medium-distance migrants and are uncommon at most migration sites. Peak spring flights occur from early March to early April, varying somewhat between years and specific sites. Dinosaur Ridge is the best site in spring to see Ferruginous Hawks. Bountiful Peak is the best fall site for Ferruginous Hawks, with close-up views common throughout September and October. Most other sites in the western United States see only a few Ferruginous Hawks each season.

Rough-legged Hawks, which breed in arctic North America from Alaska through Canada, occur in significant concentrations along the Great Lakes during spring and fall but are uncommon elsewhere on migration, especially in the southeastern United States. High numbers of Rough-legged Hawks are tallied at Whitefish Point in spring from late March to late April and at Hawk Ridge, Holiday Beach, and Lake Erie Metropark in fall from late October to late November (see tables 2 and 3). Rough-legged Hawks can be seen in winter throughout most of the West but are uncommon in the southernmost western areas from California to Texas.

Red-shouldered Hawk (*Buteo lineatus*)

Juvenile

Juvenile Eastern Red-shouldered Hawks (*B. l. lineatus*) are buff colored underneath with varying amounts of dark streaking, which is usually evenly dispersed throughout the body. **Eastern Red-shouldered Hawks are extremely similar to juvenile Broad-winged Hawks underneath; however, the streaking on Broad-winged Hawks is typically heaviest toward the sides of the breast.** Juvenile Red-shouldered Hawks may show a rufous wash on the underwing coverts, which can fade by spring. The tail has indistinct banding throughout with a wider, dark subterminal band, but the subterminal band is not always boldly defined like that of juvenile Broad-winged Hawks. Unlike Broad-winged Hawks, **all Red-shouldered Hawks have translucent comma-shaped wing panels throughout the primaries.** These panels are obvious most times from below but can be inconspicuous in poor lighting. Juvenile California Red-shouldered Hawks (*B. l. elegans*) have a mix of rufous brown barring and streaking underneath that, when densely marked, can appear similar in the field to the plumage of adults. Juvenile California birds appear longer tailed than Eastern juveniles.

The upperside of juvenile Eastern Red-shouldered Hawks is brown; some juveniles have rufous "shoulder" patches, similar to those of adults, but they are less distinct. **The pale crescent-shaped wing panels are obvious from above.** The tail is indistinctly banded, similar to that of juvenile Red-tailed and Broad-winged Hawks, but may have hints of rufous throughout. Like most light morph buteos, Red-shouldered Hawks have varying amounts of pale mottling along the upperwing coverts, and they show pale uppertail coverts. The topside of juvenile California birds is similar to that of adult Eastern birds, showing black and white banded flight feathers and rufous "shoulders."

Adult

Adult Red-shouldered Hawks are a vibrant orange underneath with faint white barring and black streaking on the body; **sometimes they appear orange overall at a distance.** The black streaking on some individuals is obvious from close up. The flight feathers are boldly banded black and white throughout. **The tail of Red-shouldered Hawks has several broad black bands with narrow white bands in between.** The topside of Red-shouldered Hawks is striking, with boldly banded black and white flight feathers, a brownish back and upperwing coverts, and contrasting rufous "shoulder" patches. California adults are slightly paler and more vibrant underneath than Eastern adults.

Red-shouldered Hawk

RS 01 - Red-shouldered Hawk, juvenile Eastern. Pale below and lightly streaked across chest (left, MI) or moderately streaked throughout body (right, MI). Note unmarked underwing coverts and translucent comma-shaped windows across primaries that all Red-shouldered Hawks exhibit.

RS 02 - Red-shouldered Hawk, juvenile Eastern (NJ). Eastern birds are brown on top with pale comma-shaped wing panels and pale mottling along upperwing coverts.

RS 03 - Red-shouldered Hawk, adult Eastern. Adults are orange underneath with black and white banding throughout remiges and tail (left, MI). Adults are boldly banded black and white on top (right, PA) with brilliant rufous "shoulders." Adult California birds are similar above but can be slightly paler overall.

RS 04 - Red-shouldered Hawk, juvenile California (CA). Juveniles show a mix of rufous streaking and barring underneath (left). Upperside (right) is adult-like but with darker head, back, and "shoulder" patches. Photo on left © Chris Neri; photo on right © William S. Clark

RS 05 - Red-shouldered Hawk, adult California (CA). Similar to Eastern adults but more vibrant underneath. © Brian K. Wheeler

Broad-winged Hawk (*Buteo platypterus*)

Juvenile

Juvenile Broad-winged Hawks are buff colored underneath with dark streaking that is typically limited to the sides of the breast. The streaking on juveniles is variable and can be light to heavy throughout the body. **Some juveniles appear barred underneath, similar to adults.** The underwing coverts of juveniles are essentially unmarked, and the underside of the tail shows faint, narrow bands with a wider, dark subterminal band. The topside of juveniles is dark brown with sparse to moderate pale mottling along the upperwing coverts.

Adult

Adult Broad-winged Hawks are whitish underneath with rufous barring on the chest (and sometimes belly) and unmarked undertail coverts. Rarely, the barring of adults is faint and inconspicuous in the field. The underwing coverts are whitish with faint rufous mottling. Some adults are solid rufous on the upper breast, appearing to have a "bib" similar to that of adult light morph Swainson's Hawks. Birds in their first year of adulthood are usually identical to older adults, but a few individuals retain their juvenile body plumage until the following year. All adults have a defined dark trailing edge to the wings, which juveniles lack. Adults are dark brown on top, with some individuals having slightly contrasting blackish hands. **The tail of Adult Broad-winged Hawks has two to three black and white bands of equal width.**

Dark morph Broad-winged Hawks occur but are rare. Juveniles are completely brown or heavily streaked throughout the body and underwing coverts, whereas adults are solid blackish on the body. The pattern on the tail and remiges of dark birds is identical to that of light birds, but the upperwing of dark morph birds is darker overall and less mottled. **Juvenile dark morph Broad-winged Hawks may appear to have a dark terminal band on the wings similar to that of adults, but it is less boldly defined.**

Broad-winged Hawk

BW 01 - Broad-winged Hawk, juvenile (MI). Typical plumage; pale below with streaking on sides of breast, and sometimes throughout belly, and relatively unmarked underwings. © Michael Shupe

BW 02 - Broad-winged Hawk, juvenile (CT). Juveniles can be virtually unmarked below. Pale wing panels are often indistinct. © Jim Zipp

BW 03 - Broad-winged Hawk, juvenile (MI). Some birds are somewhat barred, similar to adults. Note lack of dark border on wings. © Michael Shupe

BW 04 - Broad-winged Hawk, adult (MI). Adults have barred rufous chest, boldly banded tail, and dark border along wings. © Michael Shupe

BW 05 - Broad-winged Hawk, adult (NV). Barring on chest of this adult is inconspicuous. Note banded tail and dark border on wings. © Sarah Frey

BW 06 - Broad-winged Hawk, dark morph. Dark morphs appear solidly dark underneath with pale flight feathers and stocky, pointed wings. Juveniles (left, CO) have dark, smudgy border on wings and dark tail tip. Adults (right, MN) have distinct dark border on wings and boldly banded tail. Both photos © Brian K. Wheeler

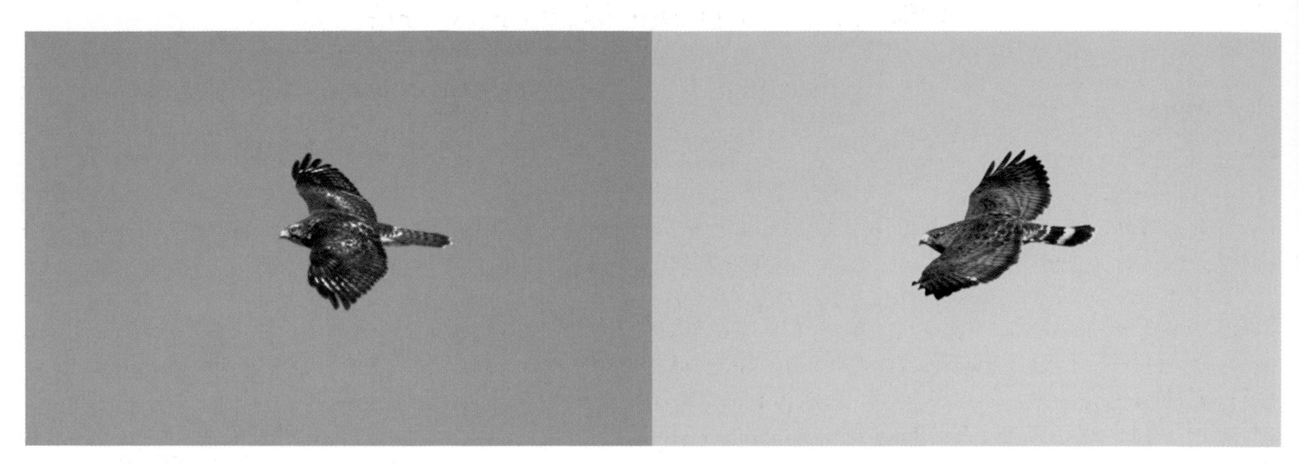

BW 07 - Broad-winged Hawk (NV). Juveniles (left) are brownish on top with pale uppertail coverts and pale mottling along upperwings. Adults (right) are dark with grayish cast to remiges and distinctly banded tail. At eye level, Broad-winged Hawks are stocky like accipiters but show broader wings and shorter tail.

BW 08 - Broad-winged Hawk, juvenile. In spring, missing primaries caused by molt on this Broad-winged Hawk (right, MI) appear similar to crescents in wings shown by **Red-shouldered Hawks** (left, MI), but they are prominent and lack a border. Birds shown same size. Photo on right © Michael Shupe

Swainson's Hawk (*Buteo swainsoni*)

Light Morph

Light morph Swainson's Hawks are whitish below with contrasting black flight feathers; **they appear two-toned underneath, or "black in back."** Juveniles have dark streaking that is limited to the sides of the breast; in rare cases, the entire breast is streaked and appears as a "bib." Swainson's Hawks lack the pale primary wing panels present on most other juvenile buteos. In fall, juveniles typically show a pale rufous wash on the body. Adult Swainson's Hawks have a dark bib that is apparent at considerable distances. Males typically have grayish cheeks, a pale rufous bib, and lack significant mottling on the body; females often have brownish cheeks and bib and rufous mottling on the belly. Be careful; **the flight feathers of juvenile birds are somewhat paler than those of adults, making juveniles less strikingly two-toned underneath, especially when backlit.** The tail of all Swainson's Hawks is dark overall with indistinct grayish banding, but it appears somewhat pale underneath with a dark tip. The dark tail tip is broader and more obvious on adults than on juveniles.

Unlike other buteos, which acquire their adult plumage after their first molt, **Swainson's Hawks acquire a sub-adult plumage.** This plumage consists of new adult flight feathers and retained juvenile body plumage. The adult remiges have a broad, dark trailing edge on the wing as in other adult buteos, but this may be difficult to see since the remiges of Swainson's Hawks are dark overall. **Juvenile and sub-adult birds often have a whitish head in spring because of fading; adults always have a dark head.** Some juveniles have a pale head even in fresh plumage. In fall, Swainson's Hawks that show wing or tail molt but that exhibit juvenile body plumage are sub-adult birds.

All Swainson's Hawks exhibit two-toned uppersides, with the blackish flight feathers contrasting with the paler brown upperwing coverts. **This two-toned appearance can be seen at great distances and is an excellent field mark for telling Swainson's Hawks from other buteos.** In spring, birds may look even more contrasting on top because of faded upperwing coverts. In poor light, however, Swainson's Hawks can appear blackish overall on top. Be aware that many adult Broad-winged Hawks show a somewhat two-toned upperside, but this contrast is between the pale wing base and dark hands. Adult Swainson's have uniform upperwing coverts, whereas the upperwing coverts of juveniles have pale tips and appear mottled. In spring, the pale tips on the upperwing coverts of juveniles may be worn, and absent as a result. In spring, juvenile and sub-adult birds are difficult to differentiate from each other from above. Sub-adults have darker flight feathers and an adult-like tail with a broad, dark tip.

Intermediate and Dark Morphs

Juvenile intermediate morph birds are moderately to heavily streaked on the underbody but typically lack extensive mottling on the underwing coverts. Adult intermediate morphs are solid rufous brown on the body but show pale underwing coverts. Juvenile dark morphs are heavily mottled throughout the underwing coverts and body but often show a slight contrast between the underwing coverts and darker remiges. Rarely, dark juveniles are completely dark on the body. Adult dark morphs are uniformly dark rufous brown or brown on the body with slightly paler rufous underwing coverts. **Dark morphs of other buteos show a two-toned appearance underneath, with a dark body and underwing coverts contrasting against paler flight feathers.** Almost all intermediate and dark-morph Swainson's Hawks have pale undertail coverts, but some birds have heavily barred undertail coverts that can appear dark in the field. The topside of dark morph Swainson's Hawks is similar to that of light morphs but is often darker overall and less two-toned. Intermediate, and especially dark morph, birds can show dark uppertail coverts.

Swainson's Hawk

SW 01 - Swainson's Hawk, juvenile (UT). Light morph juveniles (left) are pale below with dark flight feathers and dark markings typically limited to sides of chest. Intermediate morph juveniles (right) have streaking throughout chest. In fall, juveniles show a rufous tint to body that fades to whitish by spring.

SW 02 - Swainson's Hawk (UT). Dark morph juveniles (left) are streaked throughout or completely dark on body but still appear two-toned overall. All sub-adults, such as this intermediate bird (right), show juvenile-like plumage with adult flight feathers (note broad, subterminal tail band). Non-adults in spring have pale head.

SW 03 - Swainson's Hawk, juvenile light morph (UT). Flight feathers of juveniles (left) are paler than those of adults, giving them a less striking two-toned underside, especially when backlit (right). Photo on left © Sherry Liguori

SW 04 - Swainson's Hawk, adult light morph. Adult light males (left, WA) typically have whitish body, grayish cheeks, and pale rufous bib. Females (right, UT) have dark chestnut head and bib and often show sparse mottling on belly.

SW 05 - Swainson's Hawk, adult intermediate morph (UT). Note solid bib with heavily barred belly. Cheek and bib color of males (left) and females (right) are similar to those of light morphs.

SW 06 - Swainson's Hawk, adult dark morph (CA). Dark morphs are solidly dark overall with slightly paler underwing coverts.

SW 07 - Swainson's Hawk, juvenile (UT). Juveniles are two-toned above with pale mottling along upperwing coverts.

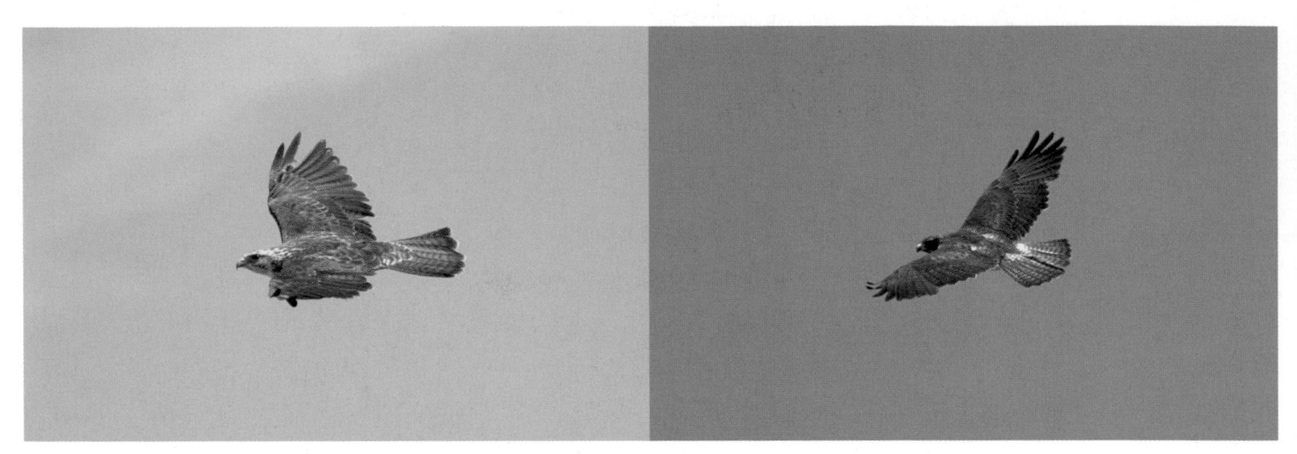

SW 08 - Swainson's Hawk (UT). Sub-adults (left, same bird as SW 02) and adults (right) are identical above, except in spring when sub-adults show pale head, such as this. Note dark uppertail coverts of intermediate morph (left) compared with pale uppertail coverts of light morph. All Swainson's Hawks show two-toned upperwings.

SW 09 - Swainson's Hawk, adult dark morph (UT). Dark (and intermediate) morph birds often show dark uppertail coverts (left). At a distance, note two-toned upperwings that all Swainson's Hawks show (right).

Red-tailed Hawk (*Buteo jamaicensis*)

The Red-tailed Hawk is perhaps the most variably plumaged raptor in North America. There are three recognized colorations, or morphs, of the Red-tailed Hawk—light, intermediate (rufous), and dark—yet **countless plumage variations, ranging from almost completely whitish to completely blackish underneath, occur within these morphs.** There are five races of the Red-tailed Hawk in the United States and Canada—Eastern (*B. j. borealis*), Western (*B. j. calurus*), Harlan's (*B. j. harlani*), Fuertes (*B. j. fuertesi*), and Florida (*B. j. umbrinus*)—but they are not always distinguishable in the field. The Fuertes and Florida races are not discussed here since they are generally nonmigratory and geographically localized.

The Eastern race of the Red-tailed Hawk occurs only as a light morph. The Western and Harlan's races occur as light, intermediate, or dark morphs. Western Red-tailed Hawks breed with both Harlan's and Eastern birds where their ranges overlap, and morphs within each race interbreed as well. As with most raptors, offspring of birds from two separate races or morphs (intergrades) can mirror a single parent or take on characteristics of both parents. **Intergrades that inherit traits of both parents are extremely difficult to specifically categorize.** Although pairings between light morph and dark morph Red-tailed Hawks are quite common, the mating of two dark birds is much less common. Western Red-tailed Hawks average slightly longer wings than Eastern Red-tailed Hawks. Therefore, Western juveniles can look quite lanky. Harlan's Red-tailed Hawks are somewhat stockier than other Red-tailed Hawks. Regardless, **shape differences between races are minor and extremely difficult to assess.**

The Eastern race of the Red-tailed Hawk occurs throughout eastern North America west to the foothills of the Rocky Mountains and south to northern Mexico. Krider's Red-tailed Hawk, a pale form of the Eastern race, ranges from the Dakotas to Edmonton, Canada. The Western race spans western North America from southern Alaska east to the eastern foothills of the Rocky Mountains and south to northern Mexico. The Harlan's race of the Red-tailed Hawk breeds throughout most of Alaska and parts of the Northwest Territories.

During migration, most Eastern and Western birds stay within their range, but some birds do stray well outside these boundaries. This is evident by the annual occurrence of dark morph Red-tailed Hawks at migration sites in eastern North America. Even though Red-tailed Hawks stray at times, **it is far more likely that light morph birds in the East that appear similar to Western birds are look-alikes rather than true Western birds, and vice versa.** Harlan's Red-tailed Hawks disperse from their breeding grounds in fall and inhabit a vast wintering area that stretches from British Columbia to the midwestern plains states and south to west Texas. Concentrations of wintering Harlan's are typically in areas with large agricultural fields such as the Samish and Skagit Flats in Washington, the Treasure Valley in Idaho, the Cache Valley in Utah, and the Arkansas River Valley on the Colorado plains (including the "hotspot" Lamar, CO). The best sites to see Harlan's Red-tailed Hawks during fall migration are the Goshute Mountains, Lucky Peak in Idaho, and along the Cascade Range from Canada to Oregon. In spring, Dinosaur Ridge is a good site to see northbound Harlan's.

Light Morph

Light morph Red-tailed Hawks are brown above and whitish below. Almost all light morph birds show dark patagial bars and dark streaking on the belly that forms a bellyband. Adults, excluding Harlan's, have a rufous tail; juveniles of all races have a brownish tail with multiple blackish bands. Rarely, juvenile Red-tailed Hawks possess a rufous tail similar to that of adults. Light morph Red-tailed Hawks show pale uppertail coverts that contrast with the darker tail; however, the uppertail coverts of heavily marked Western light morphs are often the same color as the tail. Be aware that Eastern and Western birds can look identical to each other. Because of this, **it is impossible to categorize many light morph Red-tailed Hawks to race based on plumage.**

Besides tail color, several characteristics differ between juvenile and adult Red-tailed Hawks. **Juveniles** have a plain brown upperside and head (some juvenile Red-tailed Hawks have purplish upperwing coverts, a trait not shown by other buteos), a faint dark trailing edge to the wings, and pale primary wing panels that are translucent from below. Juveniles may have a pale tawny upper breast in fall but are otherwise buff-white underneath. **Adults** usually show a golden nape, slate brown back with rufous hints, and a dark trailing edge to the wings. Back and nape color can be particularly helpful in aging Red-tailed Hawks when viewed from a head-on angle. Adults also show a pale rufous wash on the underside compared with the buff-white tone of juveniles. **The difference in color tone underneath between juvenile and adult Red-tailed Hawks is helpful in determining age when tail color is indiscernible.**

Typically, **Western** adults show an obvious rufous tint to the underside, broad patagial bars, a significant bellyband, mottling on the underwing coverts, a dark throat, and multiple dark tail bands. On average, Red-tailed Hawks from the Pacific Northwest are more heavily marked than other Western birds. **Eastern** adults often show a pale rufous hue underneath, a moderately marked bellyband, unmarked underwing coverts, a white throat, and a lack of tail bands. Like adults, juvenile Western Red-tailed Hawks are generally more extensively marked underneath than juvenile Eastern birds. Note that many Red-tailed Hawks from eastern Canada resemble typical Western birds.

Even though their plumages overlap, Eastern and Western Red-tailed Hawks often show a combination of traits particular to each race. For example, many Western birds with a lightly marked belly and underwing coverts exhibit broad patagial bars and a dark throat, whereas lightly marked Eastern birds often show a light throat and faint patagial bars. Yet Eastern birds that show a heavily marked bellyband often have unmarked underwing coverts; Western birds with an extensive bellyband typically show mottling on the underwing coverts. Heavily marked Western Red-tailed Hawks tend to show a more obvious rufous tone on the underside than typical Western birds.

Krider's, a pale form of the Eastern race, display a faint bellyband and patagial bars, which make them appear almost plain whitish underneath. They are pale brown above with a whitish head, although some juveniles show a pale brownish white head. Krider's tend to have extensive white mottling along the upperwing coverts. The tail of adult Krider's is pale rufous, or "pinkish," with a broad white base. Juveniles possess a pale brownish tail with dark bands throughout and a whitish base. **Some Eastern *borealis* birds can possess Krider's-like traits such as a pale head, lightly marked undersides, or a white-based tail.**

Juvenile Krider's Red-tailed Hawks are very similar in plumage to juvenile light morph Ferruginous Hawks. Albeit faint, the patagial bars and bellyband of most juvenile Krider's are distinct; juvenile Ferruginous Hawks lack these traits. Both species exhibit obvious white wing panels; however, Ferruginous Hawks have a brown head and darker brown upperwings. The tail of Ferruginous Hawks shows a white base similar to that of Krider's, but the distal half is distinctly dark. The distal half of juvenile Krider's tail is paler throughout. Also, Ferruginous Hawks always exhibit longer, narrower wings than Red-tailed Hawks. Although the breeding range of Krider's approaches the Rocky Mountains, this form has yet to be definitively recorded on migration there.

Adult light **Harlan's** are **brilliant white underneath**, lacking buff or rufous tones. They have broad, distinct streaks on the belly, lightly marked underwings, a dark brown head with a white throat, and a white line above and below the eye. Rarely, Harlan's show a finely barred bellyband. The upperside of Harlan's is dark brown. **Adult light morph Harlan's have white mottling that is often limited to the scapulars, whereas juveniles typically show extensive white mottling on the upperwing coverts, and whitish primary wing panels.** Adult Harlan's typically have grayish white tails with a broad dark or rufous tip.

Juvenile Harlan's are often bright whitish underneath with a white throat and white streaking on the head. Juvenile Harlan's have a banded tail similar to that of other races, but the tail of Harlan's can show wavy bands and rufous or gray mottling throughout. Some adult Harlan's show wavy banding on the remiges, or lack banding altogether. Red-tailed Hawks of other races have even banding to the remiges, but this trait is often difficult to judge in the field. Be aware that some Red-tailed Hawks of other races can look bright whitish at times like Harlan's, especially in spring.

Intermediate and Dark Morphs

Intermediate and dark morph Western Red-tailed Hawks are dark underneath with contrasting paler flight feathers. They are often darker overall on top, sometimes blackish, than light morph birds. The uppertail coverts of intermediate and dark birds are typically heavily barred (juveniles) or solidly dark (adults). Most intermediate and dark adults have multiple narrow dark tail bands. **Note: If a Red-tailed Hawk has pale uppertail coverts, it is a light morph.** True dark morph birds are uncommon; intermediate birds make up about 90 percent of all "dark" Western and Harlan's Red-tailed Hawks seen at western migration sites. Dark morph Western birds are more common in the Pacific Northwest than they are in other areas and tend to move easterly on migration, where they are seen annually along the Great Lakes.

Juvenile intermediate **Western** Red-tailed Hawks have a heavily mottled belly with streaking on the chest that is slightly less prominent than on the belly. However, some intermediate birds and intergrades between dark and intermediate morph birds are uniformly streaked throughout the underbody. The underwing coverts are heavily mottled as well, often masking the patagial bars. Juvenile dark Western Red-tailed Hawks have a uniformly mottled or solid brown body and underwing coverts, but it is often difficult to make the distinction between intermediate and dark birds under typical field conditions. Some solidly dark birds may have faint speckling on the belly, but this is difficult to see at any distance.

Adult intermediate and dark morph Western Red-tailed Hawks are dark brownish underneath with contrasting pale flight feathers. Intermediate birds have a dark rufous chest that contrasts slightly with the dark brown belly. Dark morph birds are uniformly dark on the body. **Even with close-up views, separating intermediate and dark morph Western Red-tailed Hawks in flight can be impossible.** Since the two morphs often appear identical in the field, I often categorize them together as "dark" birds.

Juvenile intermediate and dark **Harlan's** can be extremely difficult to tell in flight from Western Red-tailed Hawks. **Generally, Harlan's are blackish (not brown) underneath with white (not buff) mottling.** Intermediate Harlan's are heavily streaked underneath with less prominent streaking on the chest, whereas dark birds are solid blackish. Intermediate Harlan's often have a white throat and pale mottling on the head, whereas intermediate Westerns do not. These traits can be particularly helpful in identifying Harlan's when seen head-on. The remiges on juvenile Harlan's often have relatively broad, wavy bands, making the remiges appear heavily marked throughout the underside.

One of the most telling features on juvenile intermediate Harlan's is the extensive white mottling on the upperwing coverts and flight feathers which, along with the blackish upperside, gives Harlan's a striking appearance. Western Red-tailed Hawks typically show sparse buff mottling on the upperwing, or lack mottling altogether. Field marks normally associated with juvenile Harlan's, such as the banded tips of the outer primaries and the narrow, dark "spikes" on the tips of the tail feathers, are nearly impossible to see on flying birds. Be aware that some juvenile Red-tailed Hawks of other races will show banded tipped primaries and faint spikes on the tip of the tail.

Like juvenile Harlan's, adult intermediate and dark morph Harlan's exhibit black and white tones to the plumage. Dark morphs are uniformly dark on the body and underwing coverts, whereas **intermediate birds display varying amounts of white mottling on the chest—a trait Westerns lack.** A few intermediate Harlan's have somewhat pale carpals, which is not shown by "dark" Western Red-tailed Hawks. Most adult "dark" Harlan's have dark tips to their primaries, like those of Western birds. However, many adult Harlan's lack banding on the remiges, appearing "marbled" underneath instead. **"Dark" Harlan's usually have a pale grayish tail with a dark tip, similar to the tail of immature Golden Eagles or light morph Rough-legged Hawks, but the dark tail tip of Harlan's is narrow, smudgy, and less distinct.** The tail of some Harlan's, especially of intergrades between Harlan's and Western Red-tailed Hawks, shows varying amounts of rufous tones throughout. Adult Harlan's can also show a completely banded black and white tail, although this is uncommon and typically occurs on birds that are solid dark underneath. From below, the tail of any adult Red-tailed Hawk can appear whitish, but on Harlan's it often appears particularly white, and the dark tip is more apparent than the dark tail tip on Western Red-tailed Hawks. **Simply put, an adult dark Red-tailed Hawk with white mottling on the chest or lacking a reddish tail is a Harlan's.**

Pitfalls

Some light morph Red-tailed Hawks can appear dark in the field, especially adults with a distinct rufous tone to the underside and juveniles with heavily marked underwing linings and bellyband and streaking on the sides of the chest. By contrast, juvenile and adult intermediate birds can look light on the chest, similar to light morphs, especially at certain angles when the sun directly illuminates the chest.

Red-tailed Hawk

RT 01 - Red-tailed Hawk, Eastern (NJ). Typical Eastern birds show white throat, faintly marked underwing coverts, and moderately marked bellyband and patagium. Juveniles (left) have faintly banded tail. Adults (right) have "red" tail and dark terminal band on wings. Photo on left © Sherry Liguori

RT 02 - Red-tailed Hawk, adult Eastern (NJ). Birds from eastern Canada often show dark throat, bold bellyband and patagial bars, and rufous wash to chest. Bird on right is extremely heavily marked. Heavily marked Western birds would show barred underwing coverts, which this bird lacks. Photo on right © Sherry Liguori

RT 03 - Red-tailed Hawk, Krider's (TX). Krider's are extremely pale below with whitish head and faint patagial bars and bellyband. Juveniles (left) have pale brownish tail with white base and pale primary wing panels (see FH Pitfall 01). Adults (right) have white tail with rufous tip and dark terminal band on wings. Both photos © Brian K. Wheeler

RT 04 - Red-tailed Hawk, adult Eastern (NY). Some Eastern birds show Krider's-like traits, such as pale underside with faint markings, but lack white head and whitish tail (left). Others can show white tail base but otherwise possess typical Eastern traits (right). It is possible that both of these birds have Krider's lineage.

RT 05 - Red-tailed Hawk, juvenile Eastern (NJ). Juveniles are brown on top with pale mottling along upperwing coverts and pale primary wing panels (left). Some juveniles (right) have rufous tail similar to that of adults. Note juvenile-like tail bands and primary wing panels. Photo on right © Jamie Cameron

RT 06 - Red-tailed Hawk, adult. "Red" tail of adult **Eastern** birds (left, NJ) and white tail, head, and mottling on upperwing of adult **Krider's** (right, TX) can be seen at great distances in adequate light. Photo on right © Brian K. Wheeler

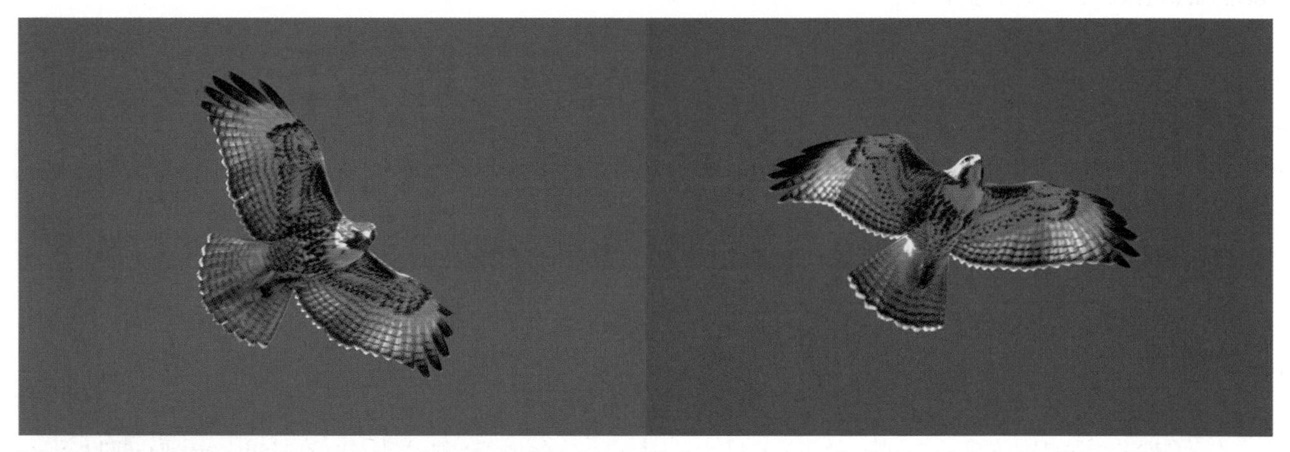

RT 07 - Red-tailed Hawk, juvenile light Western (UT). Typical light morph juvenile (left) with broad bellyband, heavily marked underwing coverts, and dark throat. Many Western juveniles have white throat and faintly marked underwing coverts (right), appearing identical to Eastern birds (see FH Pitfall 02).

RT 08 - Red-tailed Hawk, adult light Western. Typical light morph adults (left, UT) have dark throat, broad patagial bars, and faint rufous wash to underbody. Some light morph Westerns (right, NV) have strong rufous tone to underbody.

RT 09 - Red-tailed Hawk, adult light Western. Many adult Western light morphs (left, NV) have faint markings on underside similar to Eastern birds but show dark throat, unlike most lightly marked Easterns. Some Western adults (right, UT) appear identical to Eastern birds. Both photos © Sherry Liguori

RT 10 - Red-tailed Hawk, juvenile intermediate Western (NV). Typical intermediate morph (left) with streaking throughout underside but less heavily streaked on chest. Light morph birds (right) can appear similar to intermediate or dark morph birds when shadowed.

RT 11 - Red-tailed Hawk, juvenile intermediate Western (NV). When shadowed with chest illuminated (left) or when viewed head-on (right), intermediate birds can appear pale chested, similar to light morph birds, especially at a distance.

RT 12 - Red-tailed Hawk, Western. Probable light x intermediate morphs. Juvenile (left, UT) has faint streaking on chest; adult (right, NV) has dark belly and contrasting paler rufous chest, but chest is slightly paler than in typical intermediate adults and distinct patagial bars are evident.

RT 13 - Red-tailed Hawk, juvenile dark Western. Juvenile dark morphs can be uniformly heavily streaked underneath (left, UT) or solidly dark underneath (right, NV). It is possible that bird on left is an intergrade between intermediate and dark morph.

RT 14 - Red-tailed Hawk, adult intermediate Western. Even close up, contrast between dark body and slightly paler chest of adult intermediate birds (left, NV) can be difficult to discern. At a distance, intermediate birds (right, UT) often appear uniformly dark, similar to dark morph birds.

RT 15 - Red-tailed Hawk, adult dark Western (NV). Dark morph adults (left) are solidly dark on underbody; when chest is illuminated, however, it may appear to contrast with belly. Intermediate adult shown (right) under typical field conditions for comparison.

RT 16 - Red-tailed Hawk, juvenile light Western. Topside of juvenile light Western birds (left, UT) is brown overall with pale mottling along upperwing, similar to that of Eastern juveniles. Some juvenile Red-tailed Hawks (right) show a "purplish" sheen to upperwings.

RT 17 - Red-tailed Hawk, juvenile light Western. Some Western juveniles are naturally pale above with whitish uppertail coverts (left, WA), or pale in spring because of fading (right, NM). Photo on left © Sherry Liguori

RT 18 - Red-tailed Hawk, juvenile Western. Juvenile intermediate Western birds (left, UT) and especially dark morph birds (right, NV) are slightly darker above and less mottled along upperwings than light morphs. Primary wing panels of dark morph juveniles are usually faint. Note dark uppertail coverts of both birds.

RT 19 - Red-tailed Hawk, adult light Western (UT). Adult light morphs are brown above with slightly paler head and bright, rufous tails. Some Westerns (left) lack multiple tail bands, whereas others (right) have significant banding in tail. Many Western adults show rufous uppertail coverts and lack mottling along upperwing coverts.

RT 20 - Red-tailed Hawk, adult light Western (UT). Adult Western birds (left) can have particularly pale head and whitish uppertail coverts, appearing similar to Eastern birds (note lack of mottling along upperwings). Some light Westerns (right) appear extremely dark on top, identical to dark morph birds.

RT 21 - Red-tailed Hawk, light Harlan's. Typical light Harlan's are bright white underneath with distinct, broad streaks on belly, white throat, and white around eyes. Juveniles (left, NV) show banded tail similar to other races and lack dark terminal band on wings shown by adults (right, KS).

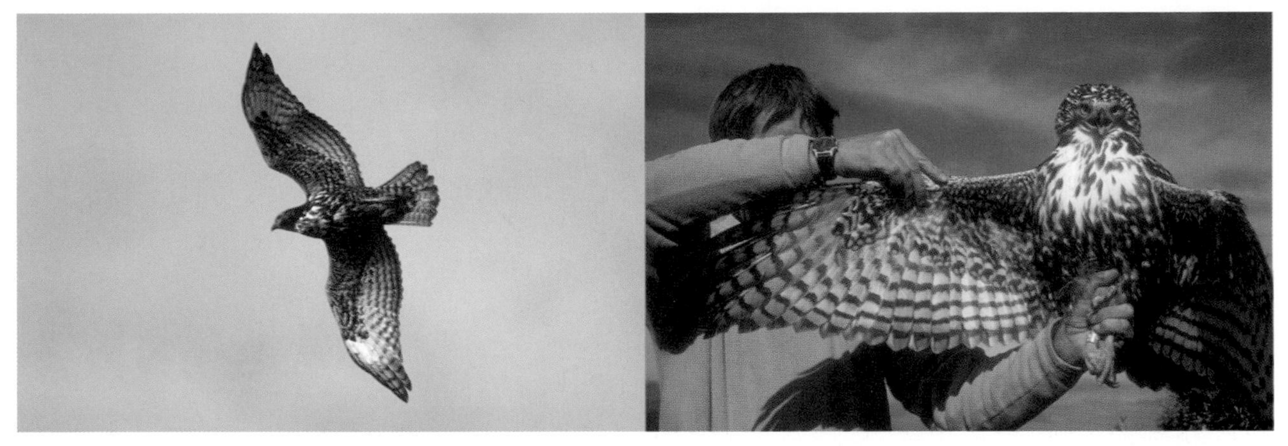

RT 22 - Red-tailed Hawk, juvenile intermediate Harlan's. Intermediate Harlan's (left, WY) are similar to Westerns but are black and white in coloration with heavily banded flight feathers. Unlike Westerns, many intermediate Harlan's (right, NV) show mottled head and white throat. Photo on left © Chris Neri

RT 23 - Red-tailed Hawk, adult intermediate Harlan's. Adult intermediate Harlan's (left, ID) are black with white, mottled chest. Note pale tail with broad dark tip. Harlan's (right, UT) with uniform mottling underneath can appear pale in direct sunlight. Note pale face, unbanded remiges, and pink-tipped tail. Photo on left © Ryan Brady

RT 24 - Red-tailed Hawk, dark Harlan's. Note solid blackish body of juveniles (left, NV) and pale forehead which dark Westerns lack. Adults (right, UT) are solid blackish and lack bands on flight feathers. Note broad, dark terminal band on wings. Photo on left © Chris Neri; photo on right © Mark Vekasy

RT 25 - Red-tailed Hawk, juvenile Harlan's. Juvenile Harlan's show whitish wing panels and bright white mottling along upperwings. Light morphs (left, ID) are brownish on top, often showing pale rufous tail. Intermediate juveniles (right, NV) are blackish with boldly banded flight feathers (same bird on right in RT 22).

RT 26 - Red-tailed Hawk, adult light Harlan's. Adult light Harlan's (left, KS) are plain brown on top with white scapulars and grayish tail with rufous tip (same bird as adult in RT 21). Light adults (right, ID) can also show grayish tail with dark tip. Note white face and scapulars.

RT 27 - Red-tailed Hawk, adult Harlan's (UT). Intermediate (left) and dark adults (right) are uniformly dark on top. Most intermediate and dark Harlan's show whitish tail with dark tip, but some dark morphs are blackish on top with dark tail. Note single albino tail feather (same bird as adult in RT 24).

RT 28 - Red-tailed Hawk, adult dark Harlan's (ID). "Partially albinistic" raptors usually show diagnostic traits. Note grayish tail with dark tip.

RT 29 - Red-tailed Hawk (NJ). Note difference in whitish tone to underside of juvenile (left) and buff tones of adult (right) light morph Red-tailed Hawks. Also note dark terminal band on wings, rufous tail, lack of primary wing panels, and slightly broader wings of adult.

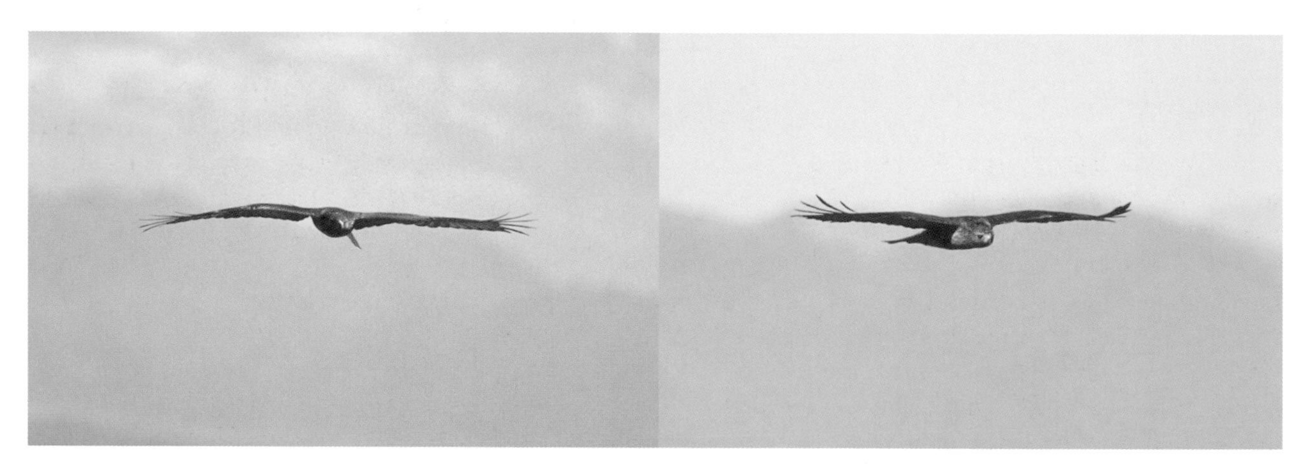

RT 30 - Red-tailed Hawk (NV). When viewed head-on, note uniform brown color of typical juvenile light morph Red-tailed Hawks (left) compared with tawny-headed appearance of adult Red-tailed Hawks (right).

RT 31 - Red-tailed Hawk, juvenile (UT). When soaring (left), juvenile Red-tailed Hawks show square pale primary wing panels. In a glide (right), outer primaries overlap each other, limiting wing panels of juveniles to inner primaries.

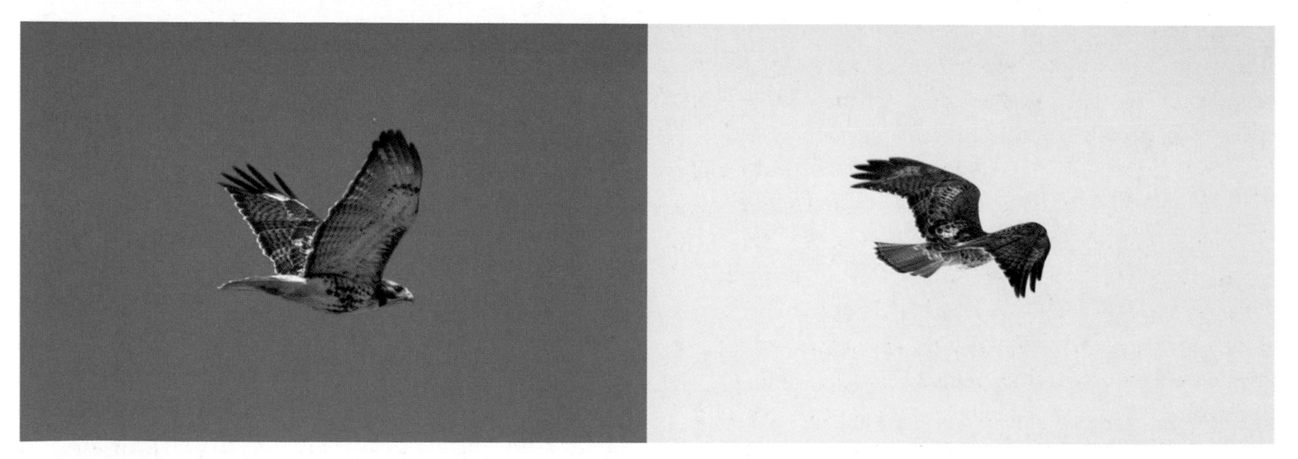

RT 32 - Red-tailed Hawk, adult. Red-tailed Hawks in their first year of adulthood (left, MI) often exhibit pale, crescent-shaped wing panels along primaries. Older adults (right, UT) may show pale wing panels, but they are muted and less obvious.

Ferruginous Hawk (*Buteo regalis*)

Light Morph

Most **juvenile** light morph Ferruginous Hawks are bright white underneath and essentially unmarked; even the black wing tips of Ferruginous Hawks are less prominent than those of other buteos. **Although indistinct, this all-white appearance alone is a good field mark to help identify Ferruginous Hawks.** With close-up views, the dark wrist commas and faint brown spotting on the underwing coverts and leggings can be seen. Be aware that some juveniles show rufous mottling on the underwings, like that of adults. The tail of juveniles appears pale from below, but the dark, faintly banded tip is apparent when fanned. All juveniles are plain brown on top with bright white primary wing panels and a smudgy brown tail with a bright white base. **These three distinct white "patches" help make identification easy, even at great distances.**

Adult light morph Ferruginous Hawks are more variable in plumage than juveniles. Some birds are predominantly whitish, similar to juveniles, but most have obvious rufous mottling on the underwing coverts. **All adults have rufous leggings that form a dark V at the belly, which can be visible at great distances.** On rare individuals, the leggings are pale and faint. Moderately marked adults typically have rusty barring on the belly. Heavily marked birds, which are less common, may have barring throughout the entire underside. Adult males are often paler overall than females, especially on the cheeks, but there is much overlap in plumage between sexes.

Adult light morph Ferruginous Hawks have a whitish tail with varying amounts of rufous or grayish mottling **but always look translucent when backlit.** Rare individuals have almost completely reddish tails like those of adult Red-tailed Hawks. The upperwing coverts of adult Ferruginous Hawks are a beautiful rusty orange that contrasts with the grayish brown secondaries. **Adults have obvious white primary wing panels like those of juveniles.** Rarely, adult light morph birds lack these panels; however, they will still exhibit rufous upperwing coverts. The head of adult Ferruginous Hawks is usually dark rusty brown but can be pale enough to appear whitish, especially when viewed head-on in direct sunlight. Ferruginous Hawks appear particularly hooded from below because of the contrast between the dark head and bright white throat and chest.

Dark Morph

Dark Ferruginous Hawks are solid brownish to burgundy brown underneath; rare individuals are blackish. All Ferruginous Hawks have bright white remiges, which lack prominent banding, distinguishing them from other dark buteos. **Juvenile and adult dark Ferruginous Hawks are extremely similar in plumage and often very difficult to tell apart in flight.** Although the dark trailing edge of the wings of adult Ferruginous Hawks is less bold than that of other buteos, it is a useful trait for telling adult Ferruginous Hawks from juveniles. In good light, most adult Ferruginous Hawks show a strong burgundy tone to the underside with a slightly darker chest. Juveniles may show this coloration underneath, but most are quite brownish with a slightly paler chest. Rarely, adults show a slightly paler chest than belly.

From above, juvenile dark morphs are plain brown with pale primary wing panels. The tail on juveniles is brownish (lacking the bright white base) with faint pale bands throughout. **Adult dark morph Ferruginous Hawks are slate brown on top with faint grayish primary wing panels; some birds exhibit a rufous tone on the upperwing coverts.** The primary wing panels on dark morphs are less obvious than those of light morph birds, but more obvious than those of other buteo species. A few adult dark morph Ferruginous Hawks have bright rufous upperwing coverts, similar to adult light morph birds, but show considerably darker uppersides overall. The topside of the tail of adults is typically plain grayish, or grayish with varying amounts of rufous mottling.

Ferruginous Hawk

FH 01 - Ferruginous Hawk, juvenile light morph (UT). Typical juveniles (left) are nearly all white underneath with dark wrist commas and faint spotting along underwing coverts and leggings. Some juveniles (right) show rufous mottling on underwings similar to adults. All juveniles have white tail with smudgy darker distal half.

FH 02 - Ferruginous Hawk, adult light morph (UT). Adults are whitish below with rufous leggings and varying amounts of mottling on underwings; tail appears whitish from below. Males (left) often have grayish face; females (right) are rufous brown overall on head.

FH 03 - Ferruginous Hawk, adult (UT). Light morph adults can show barring on belly and pale rufous wash on chest (left). Some adults are heavily barred throughout with slightly paler chest and head (right); birds such as this may be light x dark intergrades.

FH 04 - Ferruginous Hawk, juvenile light morph (UT). Juveniles are warm brown on top with bold white wing panels and white-based tail. (Same bird on right in FH 01.)

FH 05 - Ferruginous Hawk, adult light morph (UT). Typical light adults (left) show distinct white wing panels, white-based tail, and rufous upperwing coverts. Some light adults (right) show indistinct wing panels, dark rufous upperwing coverts, and mottled rufous or grayish tail.

FH 06 - Ferruginous Hawk, dark morph. Juveniles (left, UT) and adults (right, WA) are rufous brown with whitish flight feathers. Juveniles show broad, smudgy tail tip, which adults lack. Slightly paler chest on juveniles, and slightly darker chest on adults, can be impossible to see in the field. Few adults are blackish.

FH 07 - Ferruginous Hawk, dark morph. Dark morph juveniles (left, UT) are uniformly brown on top with pale wing panels. Adults (right, WA) are blackish on top with pale wing panels and pale grayish tail. Some dark adults show rufous upperwing coverts and gray-and-rufous mottled tails (same birds in FH 06).

FH Pitfall 01. Juvenile **Ferruginous Hawks** (left, UT) are similar to juvenile **Krider's Red-tailed Hawks** (right, TX), but Krider's show white head, mottling along upperwing, and paler tail overall. Krider's may also exhibit faint patagial bars and a belly-band, which Ferruginous Hawks lack. Photo on right © Brian K. Wheeler

FH Pitfall 02. Juvenile **Ferruginous Hawks** (left, UT) can be similar to some juvenile **Red-tailed Hawks** (right, WA), which can be extremely pale overall. Note smudgy tail tip, mottled wing linings, and lack of patagial bars on Ferruginous Hawk. Photo on right © Sherry Liguori

FH Pitfall 03. Dark **Ferruginous Hawks** (juvenile, left, UT) are similar to dark **Rough-legged Hawks** (juvenile, right, WA) but exhibit bright white flight feathers and lack contrasting darker carpals. Black on wing tips of Ferruginous Hawks is nearly absent.

FH Pitfall 04 (WA). Pairings between light and dark Ferruginous Hawks typically produce offspring (shown here) that are identical to one parent.

Rough-legged Hawk (*Buteo lagopus*)

Light Morph

Juvenile and adult female light morph Rough-legged Hawks are buff colored below with a blackish belly and carpal patches. **Juveniles tend to lack streaking on the chest and underwing coverts, whereas adult females typically show faint dark streaking on the chest and underwing coverts.** Some heavily marked adult females can have a dark "bib," appearing dark overall with a white "necklace" that separates the belly and bib. The tail on juveniles is white at the base with a broad, ill-defined dark tip. Adult females show a white tail with a narrower, well-defined dark subterminal band. The dark trailing edge of the wings that is noticeable on adults is less defined on juveniles.

The Rough-legged Hawk is the only buteo that exhibits true sexual dimorphism, making it possible to tell males and females apart based on plumage. The plumage of **adult males** is highly variable. Most adult males are blackish and white underneath compared to the dark brown and buff tones shown by juveniles and adult females. Some males are nearly completely mottled, lacking an obvious bellyband and carpal patches, whereas others are almost completely white with a dark bib. These pale adult males with a dark bib can be mistaken for light morph Swainson's Hawks; however, adult male Rough-legged Hawks lack dark flight feathers. **Typical males have a streaked bib, dark flanks with a paler belly, moderately streaked underwing coverts, and dark carpal patches.** Many adult males have a white-based tail with multiple narrow black bands and a broader subterminal band. The multiple tail bands of adult males are difficult to see when the tail is folded, **but the dark subterminal band of adult males is usually narrower than that of females.**

Note that adult light morph females can show a prominent bib, pale lower belly, or multiple tail bands. **When this occurs, females are often impossible to differentiate from males in flight.** Unlike males with a dark belly and bib, adult females that exhibit a prominent bib often lack extensive markings along the underwing coverts and multiple tail bands.

All light morph Rough-legged Hawks are mostly brownish on top. Birds in their first year of adulthood can exhibit pale primary wing panels similar to those of juveniles. However, they show a dark terminal band on the wings and exhibit adult plumage otherwise. Birds in their second year of adulthood with retained sub-adult outer primaries can show crescentlike wing panels from above. **Adult males usually show grayish tones throughout the back, upperwing, and remiges, whereas adult females can have grayish tones on the back only.** From above, the white base and dark tail tip of all light morph Rough-legged Hawks appear similar when folded. Light morph Rough-legged Hawks have a brownish gray head, which is sometimes particularly pale.

Dark Morph

Dark morph Rough-legged Hawks are dark underneath with pale flight feathers. In good light, **the underwing coverts are slightly paler than the carpal patches and body, especially on juveniles,** which tend to be slightly paler overall than adults. The remiges of adults are more boldly banded than those of juveniles, giving the underwings a silvery appearance at times. The trailing edge of the wings on adults shows a broad, dark band, whereas the terminal band on juveniles is smudgy and less distinct. The primary wing panels of juvenile dark birds may be less distinct than those of light birds but are visible with adequate views. The underside of the tail on dark birds appears similar to that of light birds.

The topside of dark morph Rough-legged Hawks is brownish overall, lacking pale mottling along the upperwings as in most buteos. The tail is brown with a faint black terminal band; adults often show narrow, wavy, whitish bands. From above, **the pale primary wing panels shown by juveniles are the most reliable trait for telling them from adults.** Also, many adult dark morph Rough-legged Hawks show a small white patch on the back of the head that juvenile birds lack; this trait may be helpful when aging Rough-legged Hawks from a head-on perspective. Although adult dark morph males and females can be identical in plumage, **many males are completely black on the underbody and above.** When seen in good light, black males show a dark gray-blue cast to the upperside and a boldly banded black and white tail similar to the tail pattern of adult Red-shouldered Hawks.

Rough-legged Hawk

RL 01 - Rough-legged Hawk, light morph (UT). Adult female and juvenile birds are buff below with black belly and carpal patches. Juveniles (left) have white tail with broad, dark, smudgy tip. Adult females (right) have defined dark tail tip, terminal band on wings, and mottled underwing coverts. Photo on left © Sherry Liguori

RL 02 - Rough-legged Hawk, adult female light morph (UT). Atypical females (left) can have multiple tail bands similar to adult males; they can only be seen when tail is fanned. Some females (right) have extremely heavily marked underwing coverts and bib.

RL 03 - Rough-legged Hawk, adult light morph (UT). Typical female but with pale, mottled belly similar to many adult males. Some adults show traits typical of opposite sex, making them difficult to sex.

RL 04 - Rough-legged Hawk, light morph (MI). Telling adult females (left) from juveniles (right) at a distance can be difficult. Note defined dark terminal band on wings and tail of adult female. Some adult Rough-legged Hawks, such as this, can have pale primary wing panels similar to those of juveniles.

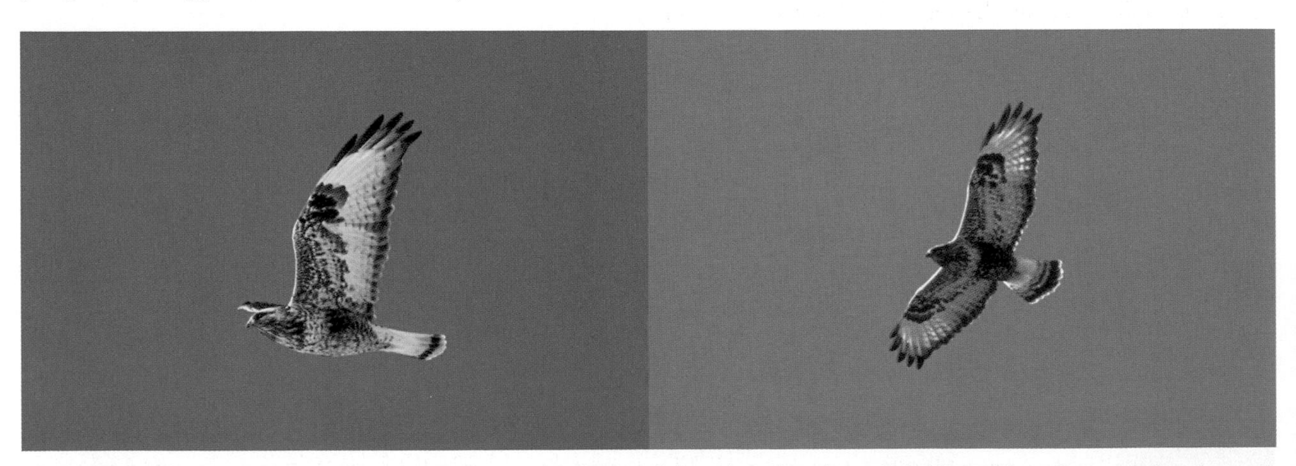

RL 05 - Rough-legged Hawk, adult male light morph (UT). Adult males often show dark bib and heavily mottled underwings. Males can have pale belly (left) or dark belly (right). Multiple tail bands shown by most males can only be seen when tail is spread.

RL 06 - Rough-legged Hawk, adult male light morph (UT). Some males (left) show dark carpal patches but pale belly and limited mottling on underwings and bib. Others (right) have dark sides to belly and flanks. Note that dark subterminal tail band is narrower than on most females.

RL 07 - Rough-legged Hawk, dark morph. Dark juveniles (left, WY) and adults (right, UT) are similar, but underwing coverts of juveniles are slightly more contrasting, and dark tail tip and terminal band on wings are less bold. Females with multiple tail bands are similar to brownish adult males. Photo on left © Chris Neri

RL 08 - Rough-legged Hawk, adult male dark morph (MI). Adult males can be similar to females, but many, like this one, are solid blackish on body and underwings. © Michael Shupe

RL 09 - Rough-legged Hawk, juvenile light morph (NV). Juvenile light morphs (left) are brownish above with pale head, white-based tail, pale wing panels, and pale mottling along upperwing coverts. Immature **Golden Eagles** (right) are similar but have longer wings, are darker overall, and lack primary wing panels.

RL 10 - Rough-legged Hawk, adult light morph (NV). Note brownish upperside with black hands of adult females (left) compared with grayish upperside of adult males (right). Adult males tend to have slightly paler head than females.

RL 11 - Rough-legged Hawk, juvenile dark morph. Dark juveniles (left, WA) are dark overall with faint mottling on upperwings and faint, grayish tail bands. Juvenile dark **Ferruginous Hawks** (right, UT) are similar but lack mottling along upperwings and show whitish primary wing panels. Photo on left © Sherry Liguori

RL 12 - Rough-legged Hawk, adult dark morph. Adult dark females and males can be brownish above with wavy tail bands; however, some females (left, UT) have brownish tail with black tip. Many males (right, NV) exhibit bluish black upperside with even, white tail bands.

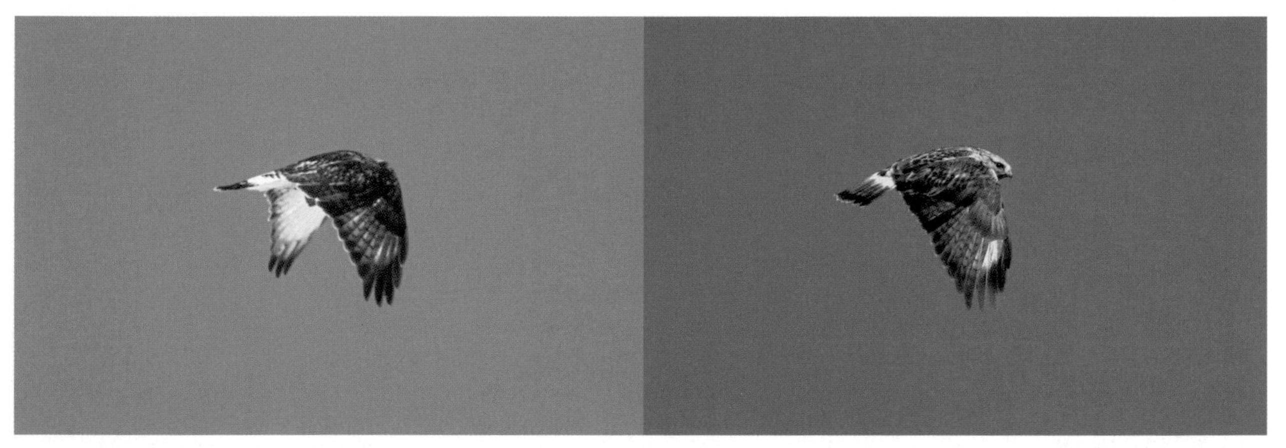

RL 13 - Rough-legged Hawk, adult light morph (UT). Rough-legged Hawks in their first year of adulthood (left) often show pale primary wing panels similar to juveniles. Birds in their second year of adulthood (right) can show crescent-shaped wing panels.

BUTEO PITFALLS

Plumage

Field marks that are typically associated with certain species of raptors are sometimes absent on that species, or present on other species. **I have seen many Red-tailed Hawks that lack an obvious bellyband and more than one juvenile Broad-winged Hawk with a distinct bellyband.** Red-tailed Hawks that lack a bellyband may appear pale overall underneath, similar to juvenile Broad-winged and Ferruginous Hawks; however, they will still exhibit patagial bars, which Broad-winged and Ferruginous Hawks lack. Be aware that adult Ferruginous Hawks can show mottling along the underwing coverts that appears similar to the patagial bars of Red-tailed Hawks.

Also be aware that specific underside field marks may be difficult to see from eye level because of the angle of the bird or shadowing, especially on dark morph birds. Most adult buteos have distinctive upperside traits, such as the rufous upperwing coverts on Ferruginous Hawks, which are helpful in wing-on identification. Juvenile buteos are more similar to each in overall color, with several species showing pale mottling along the upperwing coverts and pale primary wing panels. Regardless, the overall upperside pattern of juvenile buteos is recognizable with practice. **Almost all the buteo species herein can exhibit a pale head, pale eye-lines, pale mottling along the upperwing coverts, or dark wrist commas.** These field marks should never be used alone to identify buteos. Most buteos in spring, especially juveniles, often have a paler head than usual, which is due to fading.

Wing Panels

Adult Red-tailed and Rough-legged Hawks with retained primaries from the previous molt, which are faded and paler than the new primaries, may exhibit pale wing panels similar to those of juveniles. However, since the primaries of adults are banded throughout—unlike juvenile primaries, which are uniformly pale at the base—**the wing panels of adults are not as distinct as those of juveniles.** Birds in their first adult plumage with retained juvenile outer primaries exhibit wing panels that are similar to the crescent-shaped panels of Red-shouldered Hawks, but they are broader and more square shaped.

In spring, when juvenile buteos are undergoing primary molt, they exhibit gaps in the wings that appear similar to the crescent-shaped panels of Red-shouldered Hawks. However, these gaps lack a border and appear bolder than the "crescents" of Red-shouldered Hawks. When birds in their first adult plumage retain juvenile secondaries, which are shorter than adult secondaries, the dark trailing edge of the wings appears jagged or disjointed, unlike the even trailing edge of older buteos.

FLIGHT STYLE

When soaring, all buteos circle lazily when flying at high altitudes. At low altitudes, however, **Broad-winged Hawks soar in relatively tight circles, making quicker turns than larger buteos.** Swainson's and Ferruginous Hawks appear especially buoyant while soaring and sway from side to side at times like harriers, especially on ridge updrafts. However, the movements of Swainson's and Ferruginous Hawks are more fluid and balanced than those of harriers. **When gliding on ridge updrafts, Ferruginous Hawks hold their wings in a modified dihedral, teetering from side to side in a buoyant manner similar to that of harriers or Turkey Vultures.**

When Ferruginous, Swainson's, and Rough-legged Hawks glide off-ridge into the wind, they sway in an unsteady manner, whereas Red-shouldered, Broad-winged, and Red-tailed Hawks remain steady. Only in extremely high winds will Red-shouldered and Broad-winged Hawks become somewhat unstable, often flapping in a hurried manner to adjust their position. All buteos exhibit unwavering steadiness in a steep glide. Of the buteos, Rough-legged and Red-tailed Hawks are the most likely to hover while hunting, although Swainson's and Ferruginous Hawks do hover-hunt at times.

Wing Beat

The manner in which raptors beat their wings changes according to flight conditions. Most times, Red-tailed Hawks flap with shallow, powerful, labored wing beats. Juveniles often exhibit slightly floppier wing beats than the choppy wing beats of adults. While flapping to maintain altitude on a ridge when lift is minimal, Red-tailed Hawks will exhibit shallow, stiff, anxious wing beats in the intermittent manner of accipiters. When Red-tailed Hawks flap in order to gain altitude or accelerate forward, their wing beats are often deep, fluid, and labored.

Red-shouldered and Broad-winged Hawks flap in a quick, shallow manner that is more similar to that of accipiters than to other buteos. Compared with accipiters, the wing beats of Red-shouldered and Broad-winged Hawks are stiffer and shallower. Broad-winged Hawks flap in a distinctly choppy manner, whereas **the wing beats of Red-shouldered Hawks are somewhat "wristy" with the hands appearing to remain cupped while flapping.** The wings of Red-shouldered Hawks fall below the body when flapping, making their wing beats appear less lofty compared with Broad-winged Hawks. While soaring in light winds, Red-shouldered Hawks sometimes exhibit floppier, more labored wing beats than usual. **During courtship, all buteos display snappy, shallow wing beats similar to those of Red-shouldered Hawks.**

Swainson's Hawks (especially juveniles) and Ferruginous Hawks typically display deep, labored wing beats. The wing beats of Swainson's Hawks are fluid, similar to those of Northern Harriers, but they are stiffer and shallower compared with the loose, even tempo of harriers. Adult Swainson's often exhibit shallower, wristy wing beats, as if they were batting the wind, than juveniles. The wing beats of Ferruginous Hawks are deep, easy, and stiff along the length of the wings. The upstrokes are quick, reminiscent of Short-eared Owls but less exaggerated. The wing beats of Rough-legged Hawks are fluid and deep but stiff-handed, making the movements appear somewhat labored and wristy. Rough-legged Hawks' wing beats are typically shallower than those of Ferruginous Hawks, but **Ferruginous Hawks often display shallow wing beats when using powered flight over flat terrain.**

Size can vary greatly within Red-tailed, Ferruginous, and Rough-legged Hawks, with large females weighing more than twice as much as small males. Although the manner in which they beat their wings (i.e., shallow, lofty, stiff, etc.) remains consistent, **large Red-tailed, Ferruginous, and Rough-legged Hawks tend to flap considerably more slowly than small ones.**

SOARING

Although the buteos are similar to each other, they exhibit slight differences in shape, ranging from the relatively stocky Broad-winged Hawk to the lanky Ferruginous Hawk. Since Red-tailed Hawks are the most familiar of the buteos, they are often used as a model of comparison when discussing the differences in shape among buteos. Red-tailed Hawks exhibit long, broad wings with slightly tapered hands and a shallow arc to the trailing edge, giving the trailing edge of the wings a somewhat rounded appearance. The leading edge of the wings projects straight out from the shoulders, then bows slightly forward at the wrists. The body is broad throughout, and the head protrudes slightly less than in most other buteos. When fanned, the tail appears relatively short and broad with a shallow curve to the tip. On Red-tailed Hawks, the back edge of the wings often meets the tail beyond its base. On Red-shouldered Hawks, the tail often appears disconnected from the wings.

Red-shouldered and Broad-winged Hawks possess shorter wings than Red-tailed Hawks and lack a bulge along the trailing edge. **The wings of Broad-winged Hawks are pointed in a soar, whereas Red-shouldered Hawks have squared-off wing tips.** In comparison to Red-tailed Hawks, Broad-winged Hawks appear large headed and Red-shouldered Hawks appear somewhat long necked. When folded, the tail of Red-shouldered Hawks is long with a deeply arced tip. **The tail of Broad-winged Hawks is often square tipped when folded and particularly narrow compared to that of other buteos.** Red-shouldered Hawks bow their wings slightly forward when soaring, whereas the wings of Broad-winged Hawks exhibit a straight leading edge.

Swainson's, Ferruginous, and Rough-legged Hawks have slimmer, longer wings than Red-tailed Hawks. Swainson's Hawks have pointed wing tips similar to those of Broad-winged Hawks. Ferruginous Hawks possess long, narrow wings similar to those of Swainson's Hawks, but they are not as sharply pointed. **Ferruginous Hawks show a slight bulge at the base of the wings, appearing more angular along the trailing edge than other buteos.** Rough-legged Hawks have tapered hands but lack pointed wing tips. They also lack the angular appearance to the trailing edge of the wings shown by Ferruginous Hawks. In fact, the wings of Rough-legged Hawks project forward from the body more so than in other soaring buteos. **Male Rough-legged Hawks have slimmer wings than females and may appear more similar in shape to Northern Harriers than to other buteos.** Swainson's and Rough-legged Hawks possess a slightly narrower tail than Red-tailed Hawks, often making them appear longer tailed; Ferruginous Hawks have a long, broad tail. Rough-legged Hawks display a stockier body than Red-tailed and Swainson's Hawks, whereas Ferruginous Hawks exhibit a very broad chest and head.

Red-tailed Hawks can hold their wings flat but usually soar with a slight dihedral that starts at the shoulders and sweeps upward at the tips. Juvenile Red-tailed Hawks, which are somewhat lanky (especially Western birds), often show a more exaggerated dihedral than adults. Broad-winged and Red-shouldered Hawks typically soar on flat wings but may show slight dihedrals, particularly Red-shouldered Hawks. Swainson's and Ferruginous Hawks exhibit strong dihedrals, with Ferruginous Hawks often soaring with a slight modified dihedral. Rough-legged Hawks hold their wings in a somewhat shallow dihedral but at times display a modified dihedral in a soar.

HEAD-ON

Approaching at eye level, most buteos appear to have long wings and wide bodies. Red-tailed Hawks hold their wings in a distinctive manner, elevated slightly at the shoulders and drooped slightly at the wrists. **The wing tips of Red-tailed Hawks usually remain level with the body.** Red-tailed Hawks are broad bodied overall, and their wings appear relatively broad, long, and somewhat squared off. The white chest of light morph birds is usually obvious, although the bellyband may be obscured. Of the buteos, Red-shouldered Hawks exhibit the most severely drooped wings, which are raised slightly at the shoulders and drooped steeply along the hands. **The wing tips of Red-shouldered Hawks appear particularly squared off and are typically held below the body.**

Broad-winged and Swainson's Hawks hold their wings fairly flat with a slight droop at the hands. Both exhibit sharply pointed wing tips, which are held level with or below the body. However, Broad-winged Hawks exhibit much shorter wings than Swainson's Hawks, making them more similar in shape to accipiters than to other buteos. By contrast, the long, droopy hands of Swainson's Hawks give them an Osprey-like silhouette when approaching at eye level. However, Swainson's Hawks have shorter, less steeply bowed wings than Ospreys. Swainson's Hawks also show a more gradual bow to their wings compared with most other buteos. When viewed head-on, adult Swainson's Hawks may appear blackish because of their solid, dark upperside, but they often show a pale forehead. Remember that many buteos can have a pale head, which is most obvious on birds seen approaching at eye level.

Ferruginous Hawks exhibit extremely long wings and a broad chest from this angle. When gliding, their wings are held in a modified dihedral with wing tips held above the body in a manner similar to Turkey Vultures. Rough-legged Hawks hold their wings in a modified dihedral, but their wings appear relatively slim and long handed, similar to those of harriers. **Rough-legged Hawks appear plump at the chest and shallow at the belly. Ferruginous Hawks show a broad chest and shorter hands than Rough-legged Hawks; harriers are narrow at the chest and slim overall.**

Adult Red-tailed Hawks, especially light morphs, have a golden nape that contrasts with their dark brown back. Juvenile Red-taileds are slightly paler brown overall and lack contrast between the nape and back. Adult Red-shouldered, Broad-winged, Swainson's, Ferruginous, and Rough-legged Hawks are also darker above than their juvenile counterparts. **This difference in tone between juveniles and adults can be particularly helpful in aging buteos when viewed from a head-on angle.** Some Red-tailed

Hawks have a pale leading edge to the wings that extends from the wrist to the body; however, this field mark is evident on other raptors such as Rough-legged and Ferruginous Hawks and Golden Eagles.

GLIDING OVERHEAD

Buteos are frequently seen gliding overhead at all hawk migration sites. In a glide with wings drawn in, all buteos exhibit tapered wing tips and appear similar to each other. However, there are differences in wing shape among the buteos. **Red-tailed Hawks possess broad wings with hands that project slightly beyond the trailing edge of the wings.** Juvenile Red-tailed Hawks show a more pronounced primary projection than adults. Only in a steep glide do Red-tailed Hawks appear to have long hands. The tail of Red-tailed Hawks is relatively broad and long when folded but appears shorter than those of most other buteos. Juveniles possess a longer tail than adults, appearing more similar to other buteos.

In a glide, **Broad-winged Hawks show stocky wings with the least primary projection of the buteos.** The trailing edge of the wings is fairly straight cut, lacking a significant bulge along the secondaries. The tail of Broad-winged Hawks is the slimmest of the buteos and is typically square tipped, making it appear much like the tail of Sharp-shinned Hawks. **The tail shape of Broad-winged Hawks is an excellent trait for distinguishing them from other buteos when high overhead.** Often, Broad-winged Hawks exhibit a large-headed appearance compared with other buteos. Red-shouldered Hawks have slightly narrower wings than Red-tailed and Broad-winged Hawks. Red-shouldered Hawks exhibit relatively blunt hands that protrude farther past the trailing edge of the wings than on Broad-winged Hawks but less than on Red-tailed Hawks. The tail of Red-shouldered Hawks is wider than that of Broad-winged Hawks and has a rounded tip.

Swainson's, Ferruginous, and Rough-legged Hawks show a greater primary projection than Red-tailed Hawks. Swainson's Hawks have the slimmest wings of the buteos and exhibit the greatest primary projection. With sharply pointed wings, they may appear similar to large falcons when gliding; however, they have broader wings and a shorter tail, which does not taper toward the tip like that of large falcons. **The leading edge of the wings of Swainson's Hawks angles inward toward the body and forms a distinct M shape similar to that of Ospreys;** other buteos are less M shaped. With the wings drawn in, the remiges of juvenile Swainson's Hawks appear darker than usual, making the two-toned underside more striking.

Ferruginous and Rough-legged Hawks have broader wings than Swainson's Hawks; only in a steep glide do they exhibit an M-shaped silhouette similar to that of Swainson's Hawks. Ferruginous Hawks have broader wings and tail than Rough-legged Hawks, and they typically exhibit a square-tipped tail; the tail tip of Rough-legged Hawks is typically rounded. In a glide, the wings of Rough-legged Hawks are tapered, but the wing tips are usually more blunt than those of Ferruginous Hawks. Ferruginous and Swainson's Hawks have a fairly straight trailing edge to the base of the wings, whereas Red-tailed and Rough-legged Hawks exhibit a slight bulge along the secondaries.

Be aware that when the outer primaries of juvenile Broad-winged, Red-tailed, and Rough-legged Hawks overlap each other in a glide, the wing panels are limited to the inner half of the primaries. Since the outer primaries overlap in a glide, **adults that possess retained juvenile outer primaries will not show primary wing panels.**

WING-ON/GOING AWAY

The wings of all birds appear stockier and more pointed than usual when viewed wing-on. Red-tailed Hawks exhibit broad wings with a moderate bulge to the trailing edge and slightly projected hands. Broad-winged Hawks exhibit especially stocky wings with shorter hands and a straighter trailing edge than Red-tailed Hawks. **With stocky wings and a narrow tail, Broad-winged Hawks resemble accipiters when viewed from this angle.** However, Broad-winged Hawks exhibit sharply pointed wings, a shorter tail, and a larger head. Red-shouldered Hawks possess a slightly rounded trailing edge to the wing base, but their wings are narrower and blunt tipped compared with Broad-winged and Red-tailed Hawks.

Swainson's and Ferruginous Hawks exhibit a fairly straight trailing edge to the wing base, with a long primary projection. Rough-legged Hawks typically exhibit a rounded trailing edge with a long primary projection but slightly less pointed wing tips than Swainson's and Ferruginous Hawks. The modified dihedral of gliding Ferruginous and Rough-legged Hawks is obvious when viewed at eye level. While flapping, Broad-winged, Swainson's, and Ferruginous Hawks can show pointed wings similar to those of falcons; this is most evident from a wing-on perspective.

When headed away, buteos exhibit slimmer, straighter, more pointed wings than usual. The wings of Red-shouldered and Red-tailed Hawks remain broad, whereas Swainson's, Ferruginous, and Rough-legged Hawks show particularly narrow wings. Ferruginous Hawks exhibit a broad, "chesty" body from this angle. Broad-winged Hawks appear short-winged compared to other buteos.

Buteo Shapes

Red-shouldered Hawk | ***Broad-winged Hawk*** | ***Red-tailed Hawk***

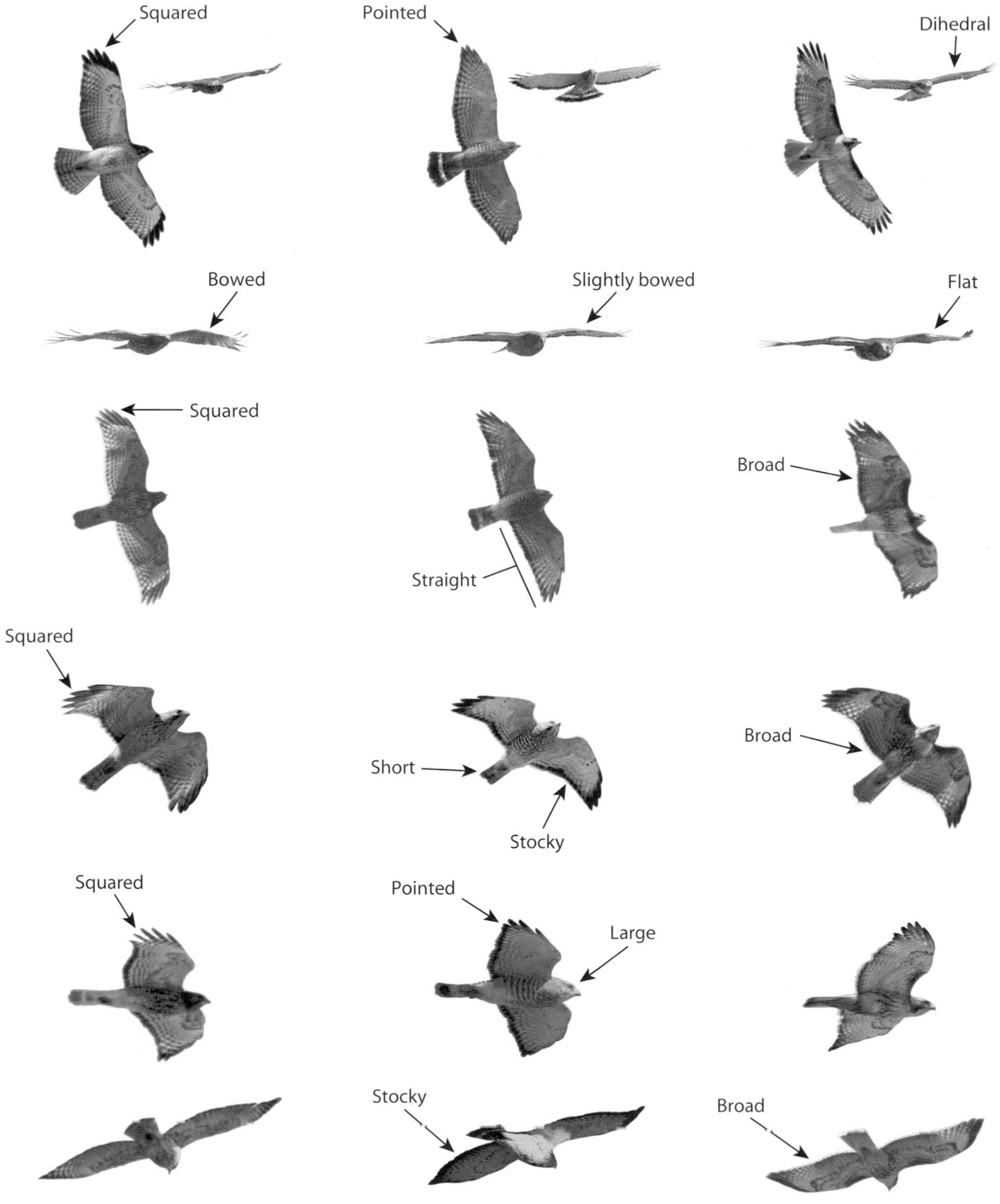

Squared
Pointed
Dihedral

Bowed
Slightly bowed
Flat

Squared
Broad
Straight

Squared
Broad
Short
Stocky

Squared
Pointed
Large

Stocky
Broad

Buteo Shapes

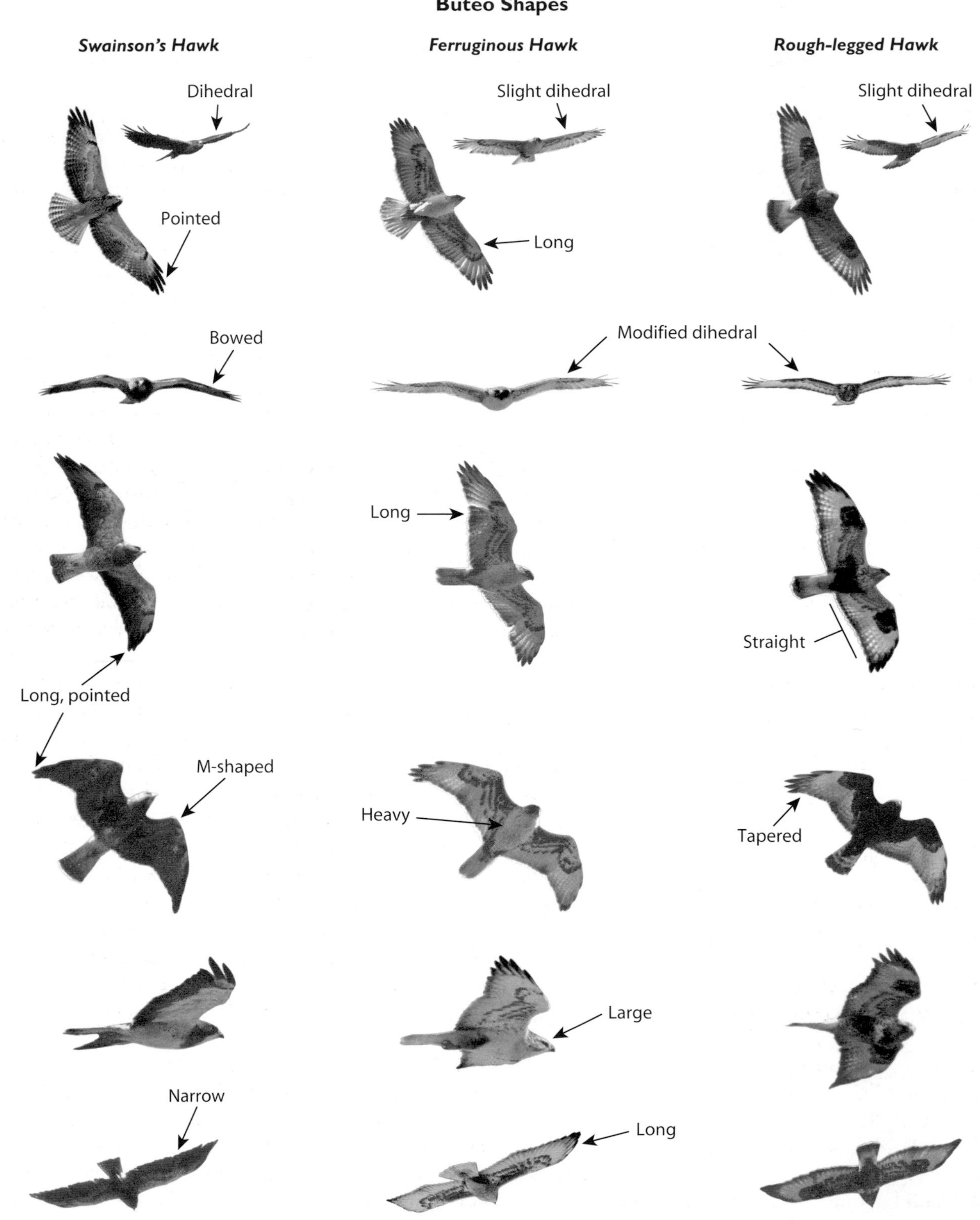

Falcons

American Kestrel, Merlin, Peregrine Falcon, Prairie Falcon

OVERVIEW

Falcons inhabit sweeping landscapes of open tundra, deserts, plains, and other vast terrains. They are known for their swift and steady flight, displaying bursts of speed rivaled by no other animal. While in direct pursuit of smaller birds, their main prey, falcons may reach speeds of more than 80 miles per hour. When stooping from above in a steep dive, Peregrine Falcons may exceed 200 miles per hour! American Kestrels are not nearly as swift as the other falcons, and often hover in midair or watch from an advantageous perch before pouncing on their main prey of mice and insects. American Kestrels, Merlins, and Peregrine Falcons occur throughout eastern and western North America. Prairie Falcons inhabit primarily the western United States, reaching portions of southern Canada and northern Mexico; records of Prairie Falcons in the East are scarce.

Size and Structure

With long, narrow, pointed wings and a long, narrow tail, falcons are built for sprinting in open country. They may hunt near densely forested or woodland habitats but do not do so within them. American Kestrels and Merlins are small, only slightly larger than jays, whereas Peregrine and Prairie Falcons are slightly larger than crows. Excluding kestrels, falcons display sexual differences in size and shape. Females can be as much as one-third larger than males. In addition, males have slimmer, more sharply tapered wings. American Kestrels and Merlins are the only falcons that display true sexual dimorphism based on plumage.

MIGRATION

Although proficient at soaring, falcons are able to fly great distances using powered flight and often do so during migration. Since they frequently catch and eat prey on the wing, they require fewer stops on migration than other raptors. Each species of falcon differs in the extent to which it migrates. Peregrine Falcons often travel long distances on migration. In fact, their name originates from the word *peregrinate,* which means "to wander." Individual peregrines have been tracked from Alaska to South America on migration, some of which have traveled to the East Coast and crossed the Atlantic Ocean from the Florida Keys. Even in strong head winds, Peregrines will not hesitate to cross large bodies of water.

American Kestrels and Merlins are short- to medium-distance migrants. Some kestrels and Merlins remain close to their breeding grounds year-round, especially birds originating from warmer climates. Birds from the northern United States and Canada may migrate considerable distances between their breeding and wintering areas. Prairie Falcons are typically short-distance migrants, with some birds remaining near their breeding territories during winter and others, especially juveniles, traveling up to several hundred miles south in fall. Much like accipiters, falcons are solitary on migration. They may appear to follow each other in pairs or small groups on days of peak flights, but they do not congregate in large flocks like Broad-winged Hawks or Turkey Vultures.

Cape May Point is the most famous site in North America for watching falcons. Autumn daily high counts of 24,875 American Kestrels, 867 Merlins, and 298 Peregrine Falcons have been recorded here. The Kiptopeke Hawk Watch, where as many as 364 Peregrines have been tallied in one day, can be equally as exciting. Curry Hammock along the Florida Keys records the highest Peregrine Falcon totals in the world, with as many as 521 seen in a day and 2,858 recorded in a season. In addition, Curry Hammock sees good numbers of Merlins each fall. High counts of Merlins are also recorded at Illinois Beach State Park, where as many as 132 have been recorded in a day. Just 60 miles north of Illinois Beach, the Concordia University Hawk Watch has tallied as many as 479 Merlins in one day.

The Golden Gate Hawk Watch, Corpus Christi, Veracruz, Hawk Ridge, Chimney Rock in New Jersey, and Fire Island in New York are other good sites to witness falcons during fall. Sandy Hook in New Jersey, the Sandia Mountains, Bountiful Peak, Braddock Bay, and Derby Hill are excellent spring sites to observe falcons. The most reliable sites for Prairie Falcons are the Manzano Mountains, Sandia Mountains, Goshute Mountains, Bountiful Peak, and Lucky Peak. Late September to early October is the peak time for falcon migration in the East; the migration can peak as much as two weeks earlier in the West. Spring falcon migration peaks from mid- to late April. Unlike other falcons, Prairie Falcons do not exhibit a distinct migration peak. However, mid-April seems to be the best time to view Prairie Falcons in spring, and early October appears to be the most reliable time in fall.

American Kestrel (*Falco sparverius*)

Male

American Kestrels are the most colorful of the falcons. Adult males are pale orange on the chest with black spotting on the belly. Juvenile males are whitish underneath with dark streaking and spotting throughout the body. The underwings of males are checkered blackish and white, appearing silvery from below and contrasting against the paler body. Also, male kestrels have white spots along the trailing edge of the wings that appear translucent when backlit; females have buff-colored spots, which are less obvious. Male kestrels have an orange tail with a broad black tip; the orange often looks whitish from below, however, when the tail is folded. When the tail is fanned, the orange coloration is obvious. The outer tail feathers of males may have multiple bands, and **when folded, the tail of male kestrels can look completely banded underneath, similar to that of a Merlin.**

Male American Kestrels have an orange back with faint dark barring and brilliant blue upperwing coverts that contrast with black remiges. In poor light, the blue upperwing coverts may look blackish, making the entire upperwing appear uniformly dark. Unlike other falcons, which acquire their adult plumage in their second year, juvenile male kestrels molt their underbody feathers during their first fall and acquire their adult plumage by late November.

Female

Female American Kestrels are pale underneath with rufous brown streaking on the body and underwing coverts, **appearing uniform in tone throughout the underside.** This trait is useful in sexing kestrels when specific plumage markings are difficult to view. Female kestrels are orange on the back and upperwings with faint dark barring throughout and blackish primaries. The tail is orange with multiple narrow black bands and a slightly broader subterminal band. From underneath, the tail appears whitish with a dark tip, unless fanned, when the orange coloration and narrow bands are evident. Adult and juvenile females are essentially identical in plumage; however, remember that kestrels molting flight feathers during fall are adults.

American Kestrel

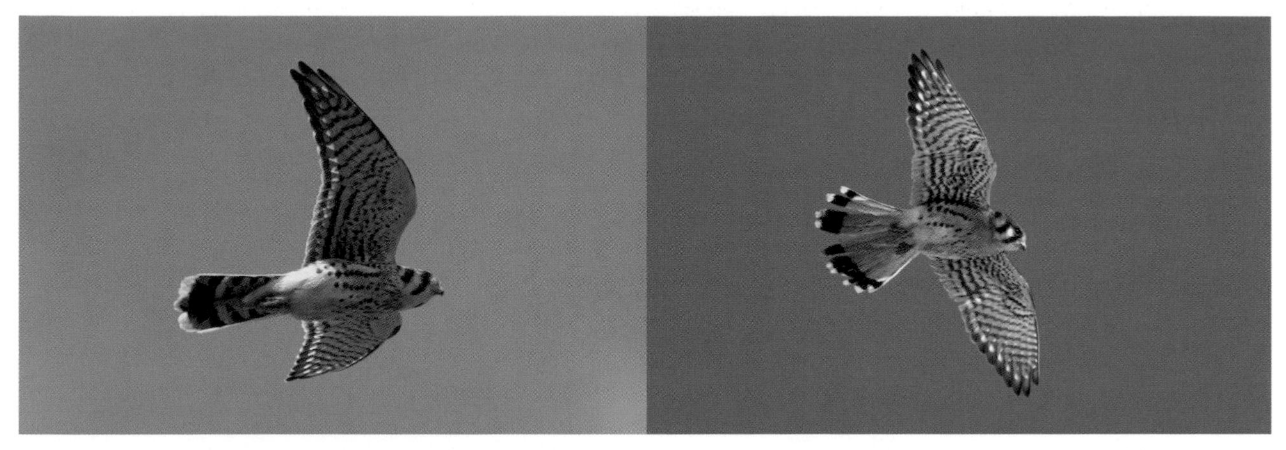

AK 01 - American Kestrel (NJ). Juveniles (left) have pale chest with dark spotting that becomes orange by late fall, as in adults (right). Note "checkered" underwings with white spots along trailing edge. Tail of males is orange with broad black subterminal band, but when folded (left) it may look banded, similar to that of a Merlin.

AK 02 - American Kestrel, female (UT). Buff below with rufous streaking on body. Tail is orange with multiple black bands. © Sherry Liguori

AK 03 - American Kestrel, male (UT). Males have orange back and tail, blue upperwings, and blackish hands.

AK 04 - American Kestrel, female (NV). Note orange upperside with black barring throughout and blackish hands.

AK 05 - American Kestrel, male (NJ). In poor light, kestrels can appear dark, similar to Merlins. Note narrow, blunt-tipped wings and white face.

AK 06 - American Kestrel. High overhead, females (left, UT) appear uniform in tone compared with silvery and buff contrasting tones of males (right, NV).

Merlin (*Falco columbarius*)

There are three distinct races of the Merlin: Taiga, or "Boreal" (*columbarius*), Prairie (*richardsonii*), and Black (*suckleyi*). All three races occur in the West, with Taiga Merlins making up about 95 percent of all Merlins recorded at western migration sites. The Taiga is the only race of the Merlin seen in the East. Prairie Merlins occur most commonly throughout the Intermountain West to the western plains states east of the Rocky Mountains and south to western Texas. Black Merlins occur predominantly along the Cascade Range on migration and along the West Coast in winter.

Taiga and Black Merlins are dark overall, excluding adult male Taigas, whereas Prairie Merlins are extremely pale. However, **each race is variable in plumage, especially the Taiga, which can be nearly as dark as Black Merlins or almost as pale as Prairie Merlins.** Intergrades between Taiga and Prairie Merlins and between Taiga and Black Merlins occur, which can make distinguishing some Merlins to race impossible.

Adult Female and Juvenile

Adult female and juvenile Taiga Merlins are buff underneath with heavy dark streaking and a pale throat. Some adult females are slightly rufous underneath with heavily marked undersides that can appear barred instead of streaked. Adult females and juveniles are brownish on top with adult females typically more slate-toned than juveniles. **Regardless of slight differences in plumage between adult female and juvenile Merlins, it is extremely difficult to tell them apart in flight.** Juvenile male Taiga Merlins may be brownish on top, or dark gray-blue similar to adult males. Merlins usually have distinct banding on the tail with a prominent white tip, a helpful field mark in distinguishing them from American Kestrels.

Prairie Merlins are considerably paler than Taiga Merlins. Juveniles and adult females are identical in plumage. Both show whitish undersides with pale rufous brown streaking, similar to the underside of female American Kestrels. Prairie Merlins and American Kestrels are also similar from above. However, **kestrels are more rufous on the upperside and show contrast between the pale upperwings and darker primaries.** The topside of juvenile and adult female Prairie Merlins also resembles the pale brown top of Prairie Falcons, but unlike Prairie Falcons, Merlins show distinct tail bands.

Juvenile and adult female Black Merlins are dark brown to blackish on top and extremely heavily streaked underneath. Black Merlins typically lack distinct tail bands and mustache markings that are shown on other races of Merlin. **Because some Taiga Merlins are extremely dark, distinguishing them from Black Merlins in flight can be impossible.**

Adult Male

Adult male Taiga Merlins are buff colored below with brown or dark rufous streaking on the body. The wrists and leggings show a yellow-gold wash, which glows brightly in direct sunlight. Adult male Taiga Merlins are blue-gray on top with slightly darker primaries. Adult male Prairie Merlins are extremely pale below with faint rufous barring on the body, resembling female American Kestrels. The topside of adult male Prairie Merlins is pale sky blue with blackish primaries. Adult male Black Merlins are buff below with heavy dark streaking, similar to female Black Merlins. The upperside of adult male Black Merlins is dark gray-blue with black primaries. Male Black Merlins typically have pale spots on the tail but lack distinct bands.

Merlin

ML 01 - Merlin, Taiga (NJ). Juveniles (left) and adult females (right) are buff underneath with dark "checkered" underwings, dark streaking on body, and dark tail with narrow white bands and white tip. In the field, adult females and juveniles are nearly impossible to tell apart.

ML 02 - Merlin, adult Taiga (NJ). Some adult females (left) show a rufous wash to underside, whereas males (right) typically show a yellowish wash to chest, leggings, and wrists. Adult males often exhibit paler underwings than adult females and juveniles.

ML 03 - Merlin, juvenile male Taiga (NV). Some juveniles are extremely rufous underneath (left). Juvenile males can show a bluish upperside similar to adult males (right) but are uniformly toned. The upperside of this individual lacks distinct tail bands and spotting on remiges as in Black Merlins, making this a possible intergrade.

ML 04 - Merlin, adult male Taiga (NV). This adult male Taiga (left) is buff toned underneath and heavily marked throughout, similar to adult females and juveniles. Note irregular tail bands (right), compared with neat, broad bluish bands of typical adult males (see ML 06). From below, this bird would be difficult to sex in flight.

ML 05 - Merlin, Black (NV). Nearly completely dark underneath. From above, Black Merlins are similar to ML 03. Adult males show bluish black uppersides.

ML 06 - Merlin, adult male Taiga (NV). Bluish uppersides with black remiges. Typical males exhibit pale face, white throat, and bluish tail bands.

ML 07 - Merlin, Taiga (NJ). Juveniles (left) and adult females (right) are dark on top with boldly banded tail. Adult females can show darker slate brown uppersides; however, they are often impossible to tell from juveniles in the field.

ML 08 - Merlin, Prairie (CO). Adult female and juvenile Merlins are pale underneath with rufous brown streaking and banded tail (left). From above (right), they are pale brownish with distinct tail bands and bold subterminal band. Photo on left © Brian K. Wheeler; photo on right © Brian Sullivan

ML 09 - Merlin, Prairie (NM). In spring, adult female and juvenile Prairie Merlins can be extremely pale overall, appearing similar to female American Kestrels. Note broad, dark subterminal tail band.

ML 10 - Merlin, adult male Prairie (UT). Adult male Prairie Merlins are whitish below with rufous streaking on body, pale yellowish leggings, and boldly banded tail. From above, males are pale bluish with darker hands, pale face, and boldly banded tail.

ML 11 - Merlin (NJ). When flapping, **American Kestrels** (left) "flip" their wings back in a hurried, fluid manner. Merlins (right) flap in a powerful, more up-and-down manner.

ML Pitfall 01. Prairie **Merlins** (left, CO) and female **kestrels** (right, NV) can appear similar to each other from below; however, Prairie Merlins are paler overall. Note stocky wings and chest of Merlin compared with kestrel. Photo on left © Brian Sullivan

ML Pitfall 02. From above, female **kestrels** (left, NV) are similar to Prairie **Merlins** (right, CO). Prairie Merlins are less rufous in tone but more uniform overall, lacking contrasting darker remiges shown by kestrels. Note Merlin has boldly banded tail with broad, dark subterminal band. Photo on right © Brian K. Wheeler

ML Pitfall 03 (NV). Adult **Sharp-shinned Hawks** (left) are bluish above, similar to adult male Merlins, but show stocky wings and lack distinct tail bands. Flapping into a headwind, **Sharp-shinned Hawks,** such as this juvenile (right), can show pointed wings. Note lightly marked, broad wings and "blobby" rufous streaking on body.

Peregrine Falcon (*Falco peregrinus*)

There are three races of the Peregrine Falcon in the United States and Canada: Tundra, or "Arctic" (*tundrius*), Anatum, or "American" (*anatum*), and Peale's (*pealei*). Tundra birds occur throughout most of North America and are the most common race seen at hawk migration sites. Anatum birds are seen across western Canada and the western United States. Peale's originate from Alaska and are most common along the West Coast in fall and winter; however, some Peale's move great distances west to east in fall and can be seen along the East Coast on occasion. All three races are similar in plumage, especially Anatum and Tundra. Distinguishing races of Peregrines in flight is often difficult, especially among adults, which are more similar to each other than juveniles. Adults are pale below with faint dark barring and bluish on top with black primaries. Juveniles are buff below with dark streaking and brownish overall on top.

Juvenile

Juvenile Peregrines are whitish below with a heavily streaked body, pale throat, and heavily barred underwings. From above, they are dark brown with slightly darker primaries and faint, pale tail bands that are typically indistinct. Most juvenile Peregrines have a prominent white tail tip like that of Merlins. Of the three races, Tundra birds are the least heavily marked underneath and most likely to show a pale forehead. Anatum birds are typically more heavily streaked than Tundra birds, show a strong rufous wash on the underside, and have a dark forehead. However, there is plumage variation in both races, and telling juvenile Tundra from Anatum birds can be difficult in flight. Peale's are extremely heavily marked underneath and above, appearing dark overall, similar to dark morph Gyrfalcons. In the East, there are several Peregrines recorded each year on migration that resemble Peale's; many of these brids, however, are of mixed origin as a result of reintroduction programs. **Peale's Peregrines are broader winged and longer tailed than Peregrines of other races, but this is difficult to judge.**

Adult

Adult Peregrine Falcons are pale underneath with faint, dark barring. Tundra and Anatum adults typically have an unmarked chest and throat, whereas Peale's may have barring throughout the underbody. Adult Peregrines have a distinct dark hood that contrasts with the pale chest. The underwings of adult Peregrines are checkered blackish and white, but adults typically appear pale overall at a distance. The topside of adult Peregrines is blue-gray with a pale blue rump and dark primaries, appearing two-toned. Anatums, and especially Peale's, can be blackish blue overall on top. The tail of adult Peregrines is dark with narrow, pale, faint bands.

Peregrine Falcon

PG 01 - Peregrine Falcon, juvenile (NJ). **Tundra** juveniles (left) are buff below with moderate streaking on body and underwings, and pale forehead. However, many birds in the East are heavily marked below (right), similar to Anatum and Peale's Peregrines.

PG 02 - Peregrine Falcon, juvenile. Juvenile **Anatums** (left, UT) are heavily marked below with a strong rufous tone on body. Anatums may or may not have pale forehead. Juvenile **Peale's** (right, NJ) are extremely dark underneath with dark head. Photo on left © Sherry Liguori

PG 03 - Peregrine Falcon, adult (NJ). Adults of all races can show a pinkish wash to body with faint barring and a white throat (left). However, adult Peale's are typically more heavily barred underneath, and some **Tundra** adults (right) are extremely whitish underneath. Plumage variation occurs in all races.

PG 04 - Peregrine Falcon, juvenile. Juveniles are brownish on top (left, NJ), or slate brown overall (right, UT). Note white tail tip but lack of distinct tail bands typical of juvenile Peregrines. Juvenile Peale's are typically dark overall on top, but variation occurs in all races.

PG 05 - Peregrine Falcon, adult. Adult **Tundra** and **Anatum Peregrines** are dark bluish on top with contrasting paler rumps (left, NJ). Some adults, especially males, are pale blue with bright blue rump and contrasting darker hands (right, NV). Adult Peale's are usually blackish blue on top.

PG 06 - Peregrine Falcon, adult (UT). Some adults, especially Anatum and Peale's, are uniformly dark blue on top.

PG 07 - Peregrine Falcon, adult (NV). Peregrines in their first year of adulthood retain varying amounts of juvenile (brown) feathers.

PG Pitfall 01. Adult **Peregrines** (left, NJ) and adult **Northern Goshawks** (right, UT) are similar, especially when headed away. Note blackish hands and level, sharply pointed wings of Peregrine compared with blackish secondaries and hands and bowed wings of goshawk.

PG Pitfall 02 (NV). Approaching at eye level, adult male **Northern Goshawks** are particularly falcon-like in shape. Note broad-based wings and white eye-line.

PG Pitfall 03 (UT). **Northern Harriers,** such as this juvenile, can appear pointed winged, but they differ considerably from falcons in plumage and flight style.

PG Pitfall 04 (UT). Some buteos, especially **Broad-winged Hawks** (such as this) and Swainson's Hawks, show pointed wings when flapping.

Prairie Falcon (*Falco mexicanus*)

Prairie Falcons are extremely pale, almost whitish, underneath with contrasting dark axillaries and dark underwing linings. Adults are lightly spotted on the underbody, whereas juveniles typically show fine, dark streaks. **Be aware that juvenile males may be spotted instead of streaked, appearing identical to adults.** Adult males are less heavily marked on the body and underwing linings than females, making them appear slightly paler overall. Prairie Falcons are pale brown above. They are paler than other North American raptors, except Prairie Merlins. The topside is identical between sexes. Juveniles are slightly darker overall than adults, especially in fall when their plumage has not yet faded. **Adults often show contrast between the body and paler tail, which juveniles lack.**

Prairie Falcon

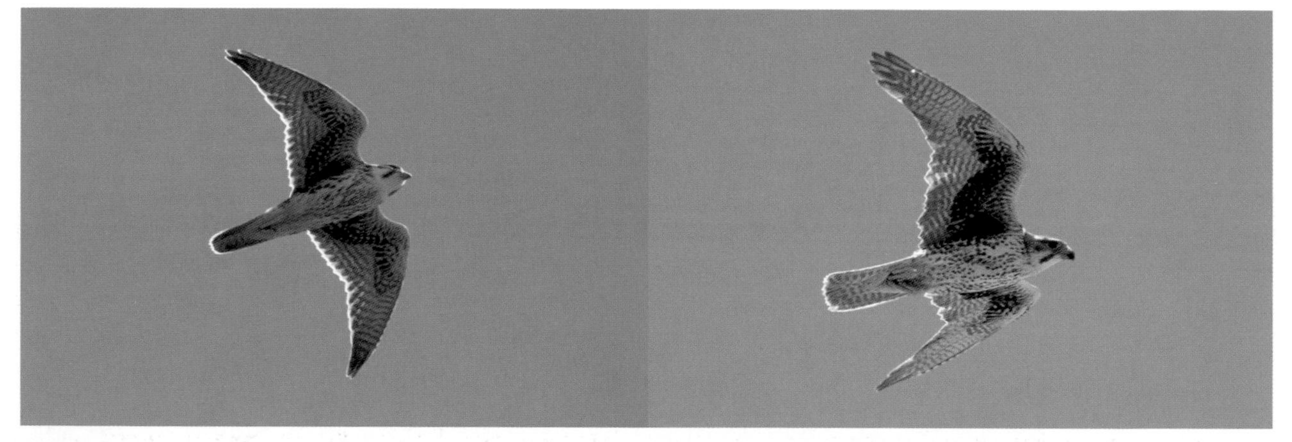

PR 01 - Prairie Falcon (UT). Prairie Falcons are pale underneath with dark axillaries and underwing coverts. Juveniles (left) are buffy with sparse streaking. Adults (female, right) are spotted as opposed to streaked on underbody.

PR 02 - Prairie Falcon. Adult males (left, UT) are less heavily marked than females. Some juvenile males (right, NM) can have short streaks on body and appear spotted like adults, making aging in flight impossible. By spring, fleshy parts can be yellow as in adults but are difficult to observe in flight. Bird on right aged and sexed in hand.

PR 03 - Prairie Falcon. Both juveniles (left, NV) and adults (right, UT) are brown above, lacking distinct markings. Adults are paler overall than juveniles, especially in fall. Note that tail of adult is typically paler than rest of upperside.

Similar Species

Gyrfalcon (*Falco rusticolus*)

Gyrfalcons are the largest of the North American falcons and the most variable in plumage, appearing almost completely white overall to completely blackish. There are three distinct color morphs: white, gray (intermediate), and dark. Adult **white morph** Gyrfalcons are nearly solid white with faint blackish markings along the upperside and black wing tips. Juveniles are similar to adults but are more heavily marked on top and faintly streaked below. In contrast, adult **dark morph** Gyrfalcons often have faint pale barring on the underside but appear solidly dark overall with no distinct field marks. Juveniles are either solidly dark or heavily streaked underneath.

Gray morph Gyrfalcons are the most common of the three morphs and the most similar in appearance to both Peregrine and Prairie Falcons. Both juvenile gray morph Gyrfalcons and juvenile Peregrine Falcons show dark streaking underneath and dark brown uppersides. **Although Gyrfalcons are thought to be darker underneath than Peregrines, many gray morph birds are less heavily marked than most Peregrines.** Adult gray Gyrfalcons and adult Peregrines are similar in plumage as well. Gray Gyrfalcons are a paler gray-blue on top than Peregrines have a pale head that lacks the distinctly hooded appearance that Peregrine Falcons show, and are lightly marked underneath, lacking contrast between the belly and chest. **A few juvenile gray Gyrfalcons may have dark mottling along the underwing linings.** When this occurs, they can be tricky to tell from Prairie Falcons, but these markings on the underwing linings are less bold than on Prairie Falcons, and the dark axillaries shown by Prairie Falcons are absent on Gyrfalcons. All Gyrfalcons have dark wing tips, a trait that only appears to show on Prairie Falcons when their wings are drawn in during a steep glide.

Male Gyrfalcons are slimmer winged than females and closely resemble Peregrines in shape, especially Peale's Peregrine, which is larger and broader-winged than other races. The wings and tail of Gyrfalcons are slightly broader than those of Peregrines, and the wing tips appear more rounded; this is most apparent on Gyrfalcons when they are soaring or flapping. Gyrfalcons have a bulky chest, and their overall shape resembles a "football with wings," whereas Peregrine Falcons are slimmer overall. The wing beats of Gyrfalcons are heavy, shallow, and less fluid or "whiplike" than those of Peregrine Falcons.

Gyrfalcon

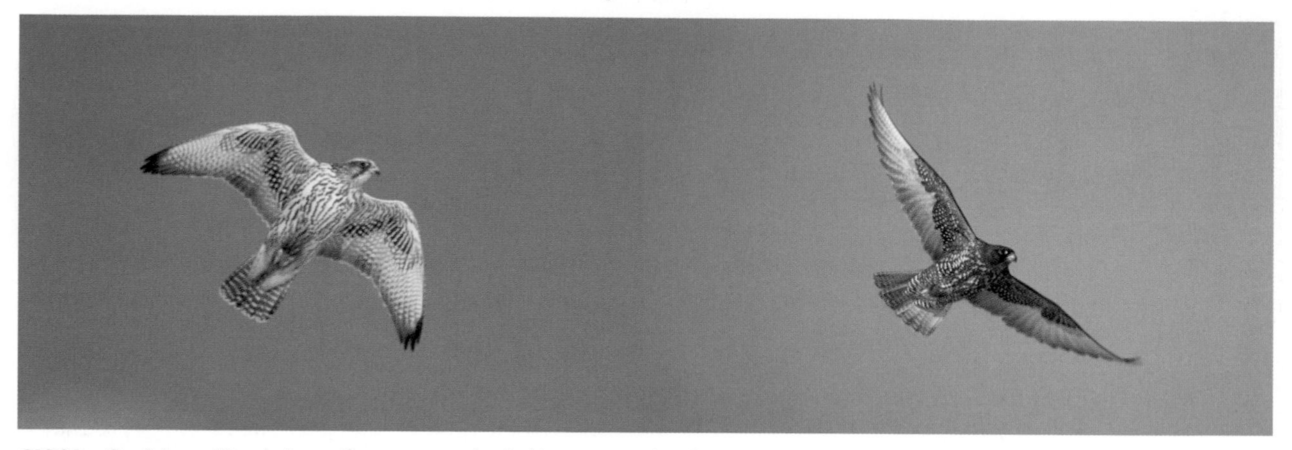

GY 01 - Gyrfalcon (Ont.). Juvenile gray morphs (left) appear paler than most juvenile Peregrines. Note atypical dark wing linings similar to Prairie Falcons, but lack of dark axillaries. Dark morphs (adult, right) appear darker underneath than Peregrines. All Gyrfalcons show pale flight feathers with dark wing tips. Both photos © Tony Beck

GY 02 - Gyrfalcon, adult gray morph. Adult gray morph **Gyrfalcons** (left, ID) are similar to adult **Peregrine Falcons** but show whitish underwings and body with dark spotting. In comparison (right), Gyrfalcon (left bird) has pale upperside and "pot-bellied" appearance as opposed to Peregrine (right bird, NJ). Shown same size.

Mississippi Kite (*Ictinia mississippiensis*)

With slim, pointed wings, Mississippi Kites can appear similar in shape to falcons, particularly Peregrine Falcons. In all postures, Mississippi Kites show slimmer wings, especially at the base, and a slimmer body than Peregrines. When gliding overhead, Mississippi Kites exhibit long hands, but they project slightly less past the base of the wings than those of Peregrines. In a soar, Mississippi Kites exhibit slim wings with relatively short hands that are somewhat blunt tipped. **The tail of Mississippi Kites is narrow but widens slightly at the tip, whereas the tail of Peregrines tapers toward the tip.** Mississippi Kites display buoyancy in flight, similar to that of Northern Harriers, and flap in an easy, gull-like manner. Peregrine Falcons are steady fliers with stiff, powerful wing beats.

Juvenile Mississippi Kites are pale underneath with heavy dark streaking. The topside is dark brown with a banded black and white tail similar to that of a Merlin. Sub-adults are grayish overall with brownish mottling on the underwing coverts, blackish remiges, and

Mississippi Kite

MK 01 - Mississippi Kite, juvenile (CO). Heavily marked rufous brown underneath (left) with dark remiges and paler head. Tail is banded black and white. Upperside (right) is dark brownish with white mottling along upperwings and paler head. Note tail widens toward tip. Both photos © Brian K. Wheeler

MK 02 - Mississippi Kite (CO). Sub-adults (left) and adults (right) are grayish below with whitish head. Sub-adults have juvenile-like banded tail (which appears all dark when folded) and often show retained dark, juvenile remiges and underwing linings. Note narrow, pointed wings similar to those of falcons. Both photos © Mike Lanzone

MK 03 - Mississippi Kite (CO). Uppersides of sub-adults (left) and adults (right) are grayish with whitish head. Sub-adults have white "upperwing bar"; adults have white secondaries and can show rusty primaries. Both photos © Mike Lanzone

a juvenile-like tail pattern. Adult Mississippi Kites are gray on top, resembling adult male Northern Harriers, but have a pale grayish white head and white secondaries that contrast against darker rust-colored primaries. They also lack the white uppertail coverts that all harriers show. The underside of adults is uniformly pale gray.

FALCON PITFALLS

Plumage

Although similar in size to American Kestrels, Merlins typically appear dark overall, whereas kestrels typically appear pale. Coloration can be helpful when telling kestrels from Merlins in flight. However, **kestrels can appear "dark" underneath if there is cloud cover or poor lighting.** One of the most common mistakes birders make in flight identification of falcons is identifying "dark" kestrels as Merlins. Remember that Prairie Merlins and adult male Taiga Merlins are quite pale underneath and can appear similar in plumage to American Kestrels.

Similar Species

From above, adult male Taiga Merlins and adult Sharp-shinned Hawks appear similar to each other. Both species are blue-gray on top and similar in size, but Sharp-shinned Hawks lack the distinct tail bands present on most Merlins. Sharp-shinned Hawks typically appear buoyant in flight and lack power, similar to American Kestrels. In contrast, Merlins fly steadily with stiff, powerful wing beats. Merlins also exhibit sharply pointed wing tips, and Sharp-shinned Hawks do not.

Adult Peregrine Falcons and adult Northern Goshawks are also similar to each other in plumage and size. Both exhibit pale undersides and bluish upperparts, but the bluish upperwing coverts of adult goshawks contrast with the blackish remiges, whereas the wings of Peregrines are blackish only along the hands. Northern Goshawks are most similar in shape to Peregrines when they are approaching at eye level. The wings of goshawks appear tapered from this angle, but they are broad at the base and less sharply pointed than those of falcons. At eye level, the white eye-line of adult Northern Goshawks is visible, whereas Peregrines lack an eye-line on their solid blackish head. The rigid, lofty wing beats of goshawks are very different from the fluid, whiplike manner of flight displayed by Peregrines.

"Pointed Wings"

Besides falcons, many raptor species can exhibit pointed wings in a glide (see Northern Harrier). The wings of accipiters and buteos can have sharply tapered hands, but the wing tips are less pointed than those of falcons. The only time falcons show blunt wing tips like those of accipiters and buteos is when they are undergoing primary molt. In addition, the wings of accipiters and buteos are broader overall with shorter hands than those of falcons. The tail of accipiters and buteos differs from that of some falcons as well, lacking a taper toward the tip like that of Peregrine and Prairie Falcons. Remember that many raptors can look pointed winged when flapping at eye level, particularly from a side view.

Wing Beat

During courtship, American Kestrels and Merlins display a manner of flight that differs from their normal flight style. During these displays, kestrels and Merlins flap in a stiff, shallow, rapid fashion with wings drooped in a "cupped" position. Both species exhibit this flight style on occasion during spring migration. Falcons frequently hunt during migration as well. While doing so, male Merlins attempt to deceive songbirds by mimicking their flight style, flicking their wings back in a flashy, rapid, intermittent fashion. Some Merlins (especially females) attempt to deceive prey by exhibiting slow, lofty wing beats before surprising the prey with a quick burst of speed.

FLIGHT STYLE

Although size can be difficult to judge in the field, American Kestrels and Merlins are sometimes easily told from the much larger Peregrine and Prairie Falcons based on size-related traits. Kestrels and Merlins display rapid wing beats and the ability to make tight, quick turns with ease. Peregrine and Prairie Falcons appear slower moving and make wide, forceful, swinging turns. When gliding high above, however, **Peregrine and Prairie Falcons are fast moving, as if they were shot out of a bow**, often passing other raptors in the sky. This is especially useful for telling high-flying Peregrines from Northern Harriers, which can show a falcon-like silhouette when gliding.

Falcons are extremely steady fliers, excluding kestrels, which appear buoyant, delicate, and butterflylike, similar to Sharp-shinned Hawks. When approaching at eye level in moderate to strong winds, **kestrels flutter like a small plane in turbulence**, whereas other falcons cut through the air in level, jetlike fashion. American Kestrels flap and glide intermittently more often than other falcons, and they rise more quickly in a soar. Their buoyant manner of flight is the most telling behavior of American Kestrels. Of the falcons, Merlins are the most likely to harass other raptors, including much larger species.

Wing Beat

Each falcon species flaps its wings in a distinctive manner. American Kestrels are the only ones that lack real power, as they flip their wings back in a sweeping, rhythmic fashion similar to the motion of a sea turtle. The wing beats of Merlins are stiff, powerful, and more perpendicular to the body, similar to those of pigeons, hence the old name of "Pigeon Hawk." Merlins seem to flap only to move forward, not to adjust their elevation or steady themselves as kestrels often do. **Even when flapping aggressively, American Kestrels move relatively slowly. When Merlins flap aggressively, they move at high speeds.**

Since Peregrine Falcons exhibit longer hands than Merlins and American Kestrels, their wing beats appear more fluid. **The hands of Peregrines appear to whip upward in a rhythmic, steady fashion similar to that of Mallards, loons, and cormorants.** Female Peale's Peregrines are larger than Peregrines of other races and exhibit floppier, more labored wing beats. Prairie Falcons flap in a similar manner to Peregrines, but their wing beats are somewhat stiffer, shallower, and quicker. The flight style of Prairie Falcons can be likened to that of a slow-beating Merlin.

SOARING

All falcons soar on flat wings, which remain pointed even when fully fanned. Among the falcons, the wings of American Kestrels are somewhat blunt tipped and the least sharply pointed. Merlins are similar to kestrels but have slightly broader-based wings than kestrels, creating an angle to the trailing edge of the wings that kestrels lack. Merlins also possess a broader chest and stockier head. **The tail of Merlins is always broader at the base, appearing shorter than that of American Kestrels.** When fanned, the boldly banded tail pattern of Merlins is obvious. The translucent spots along the trailing edge of the wings shown by American Kestrels are absent on Merlins.

Peregrine Falcons have broad-based wings with long hands compared with kestrels and Merlins. The leading edge on the wings on Peregrines is smoothly curved with a slight bend at the wrists. **The overall silhouette of Peregrines resembles that of a cocked bow and arrow**, especially when the tail is folded. Peregrine Falcons exhibit a stocky head and body compared with kestrels and Merlins. The broad tail of Peregrines tapers toward the tip when folded; the tail of kestrels and Merlins does not.

Peregrine and Prairie Falcons are extremely similar in shape. The wings of Prairie Falcons are slimmer at the hands with a slight notch where the primaries and secondaries meet. The trailing edge to the base of the wings on Peregrines is relatively straight cut, especially on adults. The pale flight feathers of Prairie Falcons often appear translucent, especially against a blue sky, giving the wings a slimmer appearance as a result. When backlit, the tail appears translucent or pinkish. Even in poor light, the dark axillaries of Prairie Falcons are almost always obvious. **The silhouette of male Prairie Falcons, which are slimmer than females, appears similar to that of American Kestrels.** However, the wings of Prairie Falcons are broader at the base and slimmer at the hands than in kestrels, which show fairly straight wings. The tail and body of Prairie Falcons are also broader than those of kestrels.

HEAD-ON

All falcons exhibit slightly drooped wings when seen at eye level. Kestrels have long wings that are smoothly curved and held hunched at the shoulders in a timid manner. In moderate to high winds on a ridge, the wing tips of American Kestrels flare upward. The wings of other falcons are more angularly curved with more sharply pointed tips, appearing more rigid than in kestrels. Merlins are stockier overall than kestrels, especially in the chest. Peregrine and Prairie Falcons show longer wings and are wider at the chest than kestrels and Merlins. Peregrine Falcons are slightly wider at the chest and have more angular wings with less blunt tips than Prairie Falcons. With adequate views, **Peregrines show a dark profile with a contrasting pale chest, whereas Prairie Falcons exhibit a pale profile overall.** The pale forehead of juvenile Tundra Peregrines is often visible at considerable distances.

GLIDING OVERHEAD

All falcons exhibit a somewhat shallow M shape in a glide. The wing tips of Merlins and Peregrines are sharply pointed, those of kestrels slightly blunt. Female Prairie Falcons often have blunt wing tips, but those of males can be sharply pointed. Kestrels and Prairie Falcons show a smooth curve to the leading edge of their wings, lacking the sharply cornered angles that Merlins and Peregrines show. The wings of kestrels and Prairie Falcons are slimmer overall than those of Merlins and Peregrine Falcons.

American Kestrels show a slimmer tail than other falcons; this trait is useful when separating falcons seen gliding overhead. Prairie Falcons typically exhibit a narrower tail than Peregrine Falcons, whereas Merlins appear to have a relatively short tail. The tail of Merlins is typically square tipped when folded, whereas the tail tip of most kestrels and Peregrines is rounded. The head of Peregrine and Prairie Falcons projects farther beyond the leading edge of the wings than that of Merlins and kestrels, and the head of Peregrine Falcons is slightly broader than that of Prairie Falcons.

WING-ON/GOING AWAY

All falcons appear similar in shape from a side angle. Peregrine and Prairie Falcons exhibit a relatively broader chest and back and longer wings than the smaller Merlins and American Kestrels. Merlins and juvenile Peregrines are similar in plumage, but Merlins are stockier overall, appearing "pot-bellied" with a slimmer, shorter tail. Merlins are also stockier overall and shorter tailed than American Kestrels, with kestrels showing blunter wing tips. When seen from above, the topside plumages of kestrels and Merlins are distinctly different and can be distinguished at considerable distances.

The difference in shape between Peregrine and Prairie Falcons is similar to the difference in shape between Merlins and American Kestrels. Prairie Falcons are slimmer overall than Peregrines and show blunter wing tips. Both Peregrine and Prairie Falcons appear extremely long winged when headed away, but the wings of Prairie Falcons appear slimmer and the tips tend to flare upward in high winds. The tail and belly of Prairie Falcons are slightly narrower than in Peregrines as well.

Falcon Shapes

Vultures, Osprey, Eagles

Black Vulture, Turkey Vulture, Osprey, Bald Eagle, Golden Eagle

OVERVIEW

Vultures, Ospreys, and eagles are large birds that appear slow moving in flight. They exhibit labored wing beats, with the exception of Black Vultures, which have relatively quick, snappy wing beats. With little practice, Turkey Vultures and Ospreys are often easy to identify, even at considerable distances. Black Vultures are fairly distinctive in flight, but their silhouette can resemble that of Bald Eagles at times. Identifying Bald and Golden Eagles in flight can be fairly easy with good views, but telling them apart under certain circumstances can be extremely difficult. Bald and Golden Eagles show several plumages before reaching adulthood at four to five years of age, and classifying immature birds to a specific age is the most challenging aspect of eagle identification.

Turkey Vultures, Ospreys, and Bald and Golden Eagles occur throughout most of North America. Black Vultures are found primarily in the eastern United States from Connecticut to parts of southern Arizona, but vagrants have been seen as far north as southern Canada and as far west as California. Turkey Vultures, Ospreys, and Bald Eagles are more common on migration in the East than in the West; Golden Eagles are more common in the West.

Size and Structure

Vultures, Ospreys, and eagles are large birds with long, broad wings. Black Vultures have relatively stocky proportions with short, broad wings and a short tail. Ospreys have long, narrow wings. Although sexual dimorphism is not apparent among vultures, Ospreys, and eagles, the difference in size between male and female eagles can be marked. Female eagles can weigh up to 14 pounds, appearing massive in flight and exhibiting extremely labored wing beats. Male eagles can weigh as little as 6 pounds. Even small eagles appear large in the field, however.

MIGRATION

Vultures and eagles tend to migrate somewhat later in fall and earlier in spring than most other raptors. Osprey migration is somewhat early in fall and late in spring, coinciding with the migration of smaller raptors such as Sharp-shinned and Broad-winged Hawks and American Kestrels. Bald Eagles and Ospreys readily cross large bodies of water, whereas vultures and Golden Eagles tend to avoid water barriers. Ospreys are seen throughout North America but are recorded along the Atlantic Coast in far greater numbers than anywhere else. They are long-distance migrants, traveling as far south as South America in fall. Juvenile Ospreys remain on their wintering grounds until their second spring, when they acquire their adult plumage. Consequently, almost all Ospreys observed in North America during spring migration are adults. Cape May Point, Kiptopeke, and Veracruz are the best sites to observe migrating Ospreys in fall. Observers at Cape May Point and Kiptopeke have recorded more than 1,000 Ospreys in a single day, with Cape May Point tallying a seasonal high count of 6,734 Ospreys in the fall of 1996 (see tables 2 and 3). Ospreys can also be seen in significant numbers along the shorelines of the Great Lakes in spring and fall. The Wasatch Mountains in spring and the Goshute Mountains in fall are the best places to see Ospreys in the West.

Unlike Ospreys, eagles are short- to medium-distance migrants and rarely travel farther south than northern Mexico. The bulk of eagle migration takes place from late March to mid-April in spring and from late October to mid-November in fall. However, a significant migration of juvenile Bald Eagles takes place in late September at many migration sites. In fall, Hawk Ridge sees the largest number of Bald Eagles of any site in North America, with as many as 743 recorded in a single day. In spring, the West Skyline Drive Hawk Watch just south of Hawk Ridge has seen up to 822 Bald Eagles in a day. Cape May,

Kiptopeke, and Hawk Mountain are also good eastern fall sites to observe Bald Eagles. In the West, Bountiful Peak in Utah and Commissary Ridge in Wyoming see the most Bald Eagles.

Franklin Mountain in New York and Hawk Mountain are good eastern fall sites to see Golden Eagles, whereas Tussey Mountain in Pennsylvania (near State College) is the premiere spring site in the East. In the West, the Bridger Mountains in Montana, Mount Lorrette in Alberta, and South Livingstone in Alberta see the most Golden Eagles of any fall site. Mount Lorrette has seen as many as 4,753 Golden Eagles in a season. Even more incredible is the 1,071 Golden Eagles tallied in a single day at South Livingstone (see tables 2 and 3). Rogers Pass in Montana sees the most Golden Eagles of any spring site.

Peak migration for vultures typically coincides with the peak time period for eagle migration. Lake Erie Metropark, Corpus Christi, and Veracruz are among the best sites to witness Turkey Vulture migration in fall. The largest spring flights of Turkey Vultures in North America are recorded at Braddock Bay and Derby Hill. Black Vultures can be seen in good numbers at Cape May Point, Kiptopeke, Corpus Christi, and Veracruz in fall. Spring migration of Black Vultures occurs along the East Coast, but significant numbers are rarely reported from any hawk migration site.

Black Vulture (*Coragyps atratus*)

Black Vultures are completely blackish underneath and above with pale silvery whitish outer primaries. In poor light, the outer primaries may appear dark and the wings may lack any contrast. Although the head of juveniles is blackish and the head of adults dark grayish, this difference in coloration is often difficult to discern in the field.

Black Vulture

BV 01 - Black Vulture (NJ). All Black Vultures are black with pale outer primaries and pale feet (left). Pale outer primaries often appear dark in the field when shadowed (right). Note stocky squared-off wings and short tail. Photo on right © Kevin Karlson

BV 02 - Black Vulture (AZ). Topside is black with pale outer primaries, similar to underside. © Chris Neri

Turkey Vulture (*Cathartes aura*)

Turkey Vultures are nearly completely blackish underneath and above. Under most conditions, **the flight feathers appear pale or silvery underneath and contrast with the body and underwing coverts, giving the underside an overall two-toned appearance.** The head of juveniles is dark with a dark bill, becoming reddish with a whitish bill on adults. Sub-adults at one to two years old typically have a pale bill with a dark tip and a partially reddish head. The head and bill color of Turkey Vultures is difficult to see in poor light or at a distance. Turkey Vultures almost never show pale mottling along the upperwing. Since vultures roost communally, however, some birds may have small white blotches on the upperside because of droppings from other roosting birds. These blotches appear irregular and inconsistent in pattern among birds.

Similar Species

Zone-tailed Hawk (*Buteo albonotatus*)

Although buteos, Zone-tailed Hawks are remarkably similar in appearance to Turkey Vultures. Zone-tailed Hawks are mostly black underneath with slightly paler flight feathers which have faint, narrow dark bands with a dark terminal band. Juvenile Zone-tailed Hawks show a less-defined terminal band on the wings than adults, and they may show pale primary wing panels similar to those of other juvenile buteos. From underneath, the tail of juveniles appears pale with a smudgy dark tip; adults have a broad dark tip with one or two distinct whitish bands at the base. When the tail of adults is folded, a single broad white band is visible at the base. The yellow feet of Zone-tailed Hawks can be visible at considerable distances.

The upperside of adult Zone-tailed Hawks is completely blackish but can show a dark grayish sheen in good light. The head is dark with a pale forehead and yellow cere that are obvious from a head-on view even at fair distances. The tail bands of adults typically appear pale grayish from above and can be difficult to see from a distance or when the tail is folded. Juveniles are somewhat plain blackish on top with faint, pale tail bands that are often difficult to see.

Zone-tailed Hawks are slightly smaller than Turkey Vultures overall but exhibit an extremely similar shape. Without careful study, individual Zone-tailed Hawks can pass by undetected, but when mixed in with Turkey Vultures they are more noticeable. Zone-tailed Hawks have a slimmer body and slightly narrower wings and tail than Turkey Vultures. **In a moderate to steep glide, the wings of Zone-tailed Hawks taper more toward the tips than those of Turkey Vultures. In a shallow glide, Zone-tailed Hawks show a straighter trailing edge on the wings, especially when headed away.** Since Turkey Vultures spend a considerable amount of time on the ground, their tail tip is often worn. The tail tip on Zone-tailed Hawks usually lacks wear.

Zone-tailed Hawks not only resemble Turkey Vultures in plumage and shape, but their flight style is extremely similar as well. Zone-tailed Hawks often soar with their tail closed, swaying in a buoyant, wobbly manner much like that of Turkey Vultures, especially on ridge updrafts. Zone-tailed Hawks glide with a modified dihedral like that of Turkey Vultures, but they exhibit slightly quicker, stiffer wing beats and maneuver with greater agility compared to Turkey Vultures.

Turkey Vulture

TV 01 - Turkey Vulture. Turkey Vultures of all ages are black underneath with silvery flight feathers. Juveniles (left, UT) have dark head and bill. Sub-adults (right, NJ) have reddish or reddish gray head with dark bill.

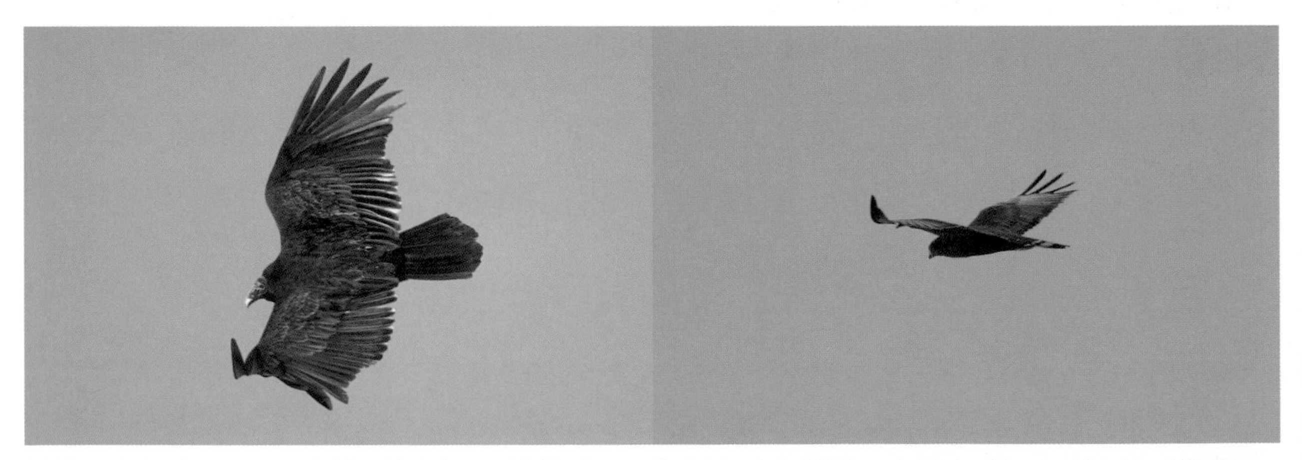

TV 02 - Turkey Vulture, adult. Topside is brownish black overall. Adults (left, NJ) have bright red head with white bill. **Zone-tailed Hawks** (right, AZ) are similar to Turkey Vultures but often show a grayish sheen on remiges, and adults (pictured) show faint grayish tail bands. Photo on right © Brian Sullivan

ZT 01 - Zone-tailed Hawk. Zone-tailed Hawks are similar in plumage to Turkey Vultures. However, juveniles (left, Mexico) have paler remiges with smudgy dark border and pale tail with dark tip. Adults (right, AZ) have dark terminal band on wings and black and white banded tail. Photo on left © Mike Lanzone; photo on right © Brian Sullivan

TV Pitfall 01 (NM). In flight, **Zone-tailed Hawks** (left) and **Turkey Vultures** (right) are extremely similar in shape and flight style. Note slightly broader wings, body, and tail and darker flight feathers of Turkey Vulture. From most angles, Zone-tailed Hawks exhibit pale forehead and larger head than vultures.

Osprey (*Pandion haliaetus*)

Ospreys are brilliant white underneath with contrasting dark carpals and dark flight feathers. Be aware that the tail of Ospreys may appear pinkish, similar to that of Red-tailed Hawks, when backlit. Juvenile and adult Ospreys are nearly identical in plumage and often difficult to tell apart in the field. Both can show a faint rufous wash on the underwing coverts, with **the wash on juveniles typically more obvious than that of adults.** Also, both juvenile and adult Ospreys can exhibit pale primary wing panels when backlit. Ospreys of all ages and sexes can show streaks on the chest that form a faint bib. **Adult females typically show a more pronounced bib than adult males, but sexing or aging Ospreys on this trait alone is often inaccurate.**

The upperside of Ospreys is blackish with a white head and dark eye-line. **Juvenile Ospreys show pale edges on the upperwing coverts, creating a scaled appearance to the topside, whereas adults are uniformly dark.** The tail of Ospreys appears dark from above, unless fully fanned when the whitish bands are evident. Adult males typically have a slightly blacker upperside than females, but this can be difficult to judge without a side-by-side comparison of the sexes. Juvenile Ospreys are more likely to show extensive streaking on the crown than adults.

Osprey

OS 01 - Osprey, juvenile. All Ospreys are bright white underneath with dark carpals and flight feathers. Note pale rufous wash on underwings and streaked bib of most juveniles (left, NJ). Juveniles that lack wash on underside (right, UT) appear identical to adults from below. Note white edging on flight feathers which adults can lack.

OS 02 - Osprey, adult (UT). Adults can have unmarked chest (left), or streaking on chest that forms a bib (right). Typically, birds with unmarked chest are males and birds with significant bib females. However, sexing Ospreys can be inaccurate. All Ospreys have white head with dark eye-line.

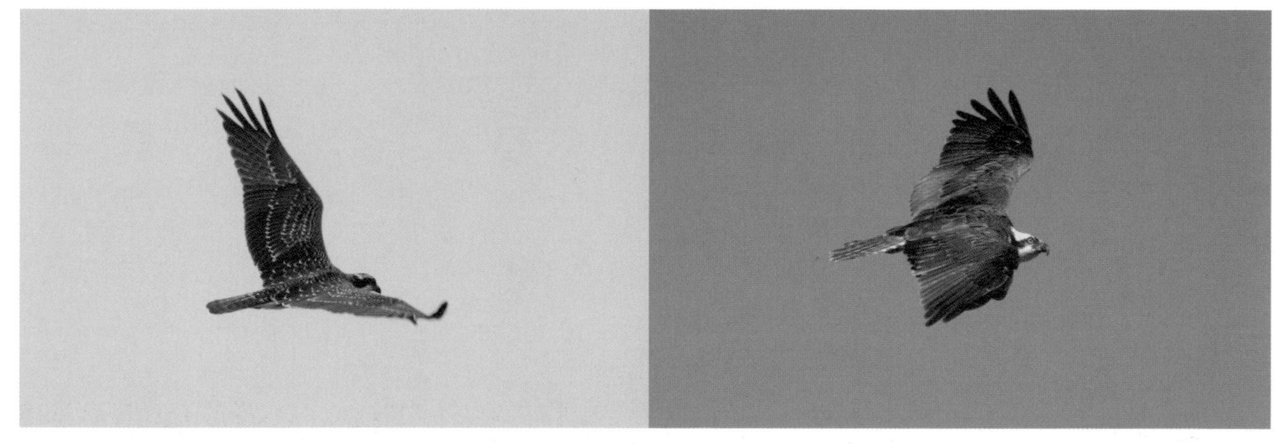

OS 03 - Osprey (NJ). Note pale feather edgings on upperside of juveniles (left) compared with uniform dark brown upperside of adults (right). Juveniles often have dark streaking on crown rather than brilliant white head of adults. Adult males tend to be blacker above than females, as shown here.

Bald Eagle (*Haliaeetus leucocephalus*)

Juvenile

Juvenile Bald Eagles are mostly blackish underneath with white axillaries and varying amounts of white along the underwing coverts and secondaries. The tail of juvenile Bald Eagles typically appears whitish from below with a smudgy, dark tip. The two innermost primaries of juvenile Bald Eagles have pale tips that are translucent when backlit. **These small translucent windows on juvenile Bald Eagles look similar to gaps in the wings caused by missing feathers.** The inner primaries of sub-adult birds may show small windows, but the primaries have dark tips that border the wings, filling in the gaps shown by juveniles. By fall, the belly of most juveniles fades from blackish to brown. **By spring, juveniles can show a pale brown to almost whitish belly, similar to the white belly of sub-adult I and II birds.** From late May through June, some juvenile Bald Eagles observed at northeastern migration sites lack feather wear or fading. These black-bellied birds are recently fledged juveniles from the southeastern United States that have dispersed north before heading back south in fall.

When viewed from above, juvenile Bald Eagles are dark overall with varying amounts of white mottling in the tail, which is most apparent when the tail is spread. By fall, the back and upperwing coverts fade to brown and contrast against the blackish flight feathers, giving the upperwing an overall two-toned appearance similar to that of Swainson's Hawks. **This two-toned pattern on the upperwing of juveniles is not shown by Bald Eagles of other age classes.** By spring, the back and upperwing coverts may be extremely faded, almost whitish.

Sub-adult I and II (White Belly I and II)

Sub-adult I Bald Eagles (1–2 years old) are similar in plumage to juveniles but have a conspicuous white belly, white back, and paler crown. The upperwings are mostly blackish, lacking the uniform fading of juveniles. Sub-adult I birds have acquired several new secondaries, which are slightly shorter than the remaining juvenile secondaries. **These sub-adult secondaries give the trailing edge of the wings an irregular appearance,** which with adequate views can be noticeable at considerable distances. As with Golden Eagles, juvenile and sub-adult I Bald Eagles have slightly broader wings than older eagles.

Sub-adult II Bald Eagles (2–3 years old) have a white belly like sub-adult I birds, or a brown belly like juveniles (because of replaced brown belly feathers). **Sub-adult II Bald Eagles with a white belly may be extremely difficult to tell from sub-adult I birds in the field.** Compared to sub-adult I birds, sub-adult II birds typically show a mostly dark back with some white mottling, and the cap is often whiter. Since sub-adult II birds typically replace the remaining juvenile secondaries that sub-adult I birds retain, they lack an irregular appearance on the trailing edge of the wings. This straight-cut trailing edge on the wings of sub-adult II Bald Eagles is sometimes the only way to differentiate them from sub-adult I birds in flight. Some sub-adult II birds, however, retain one or two juvenile secondaries, which protrude from the back edge of the wings. Also, sub-adult II, and older Bald Eagles, exhibit silvery "commas" along the base of the primaries that are pronounced below when illuminated but sometimes difficult to see otherwise. Unlike juveniles, sub-adult II Bald Eagles with a dark belly have limited white mottling on the underwings and axillaries and lack a two-toned appearance on the upperwings.

Sub-adult III (Transition)

Sub-adult III Bald Eagles (3–4 years old), or birds in transition from immature plumage to their first year of adulthood, can show significant variation. **Most transition birds have an almost completely dark body with limited white blotches on the underwings and axillaries.** The head is typically whitish with a smudgy dark eye-line similar to that of Ospreys, but it can range from almost completely white to mostly dark. Transition birds have uniformly dark uppersides but can have sparse white mottling on the back. The tail of transition birds varies from mostly dark to almost completely white but is typically white with a dark tip. **Transition Bald Eagles with a mostly dark head and a white tail with a dark tip resemble immature Golden Eagles.**

Adult

With a brilliant white head and tail that contrast with an almost black body, adult Bald Eagles are unmistakable. Although distinctive, adults can be tricky to identify under certain conditions. **In poor light, adult Bald Eagles can look uniformly dark. In adequate light, flying against a white background such as snow or cloud cover, adults can look headless.** Birds in their first year of adulthood, or Sub-adult IV birds (4–5 years old), typically have faint dark streaking on the head, a dark tail tip, and minimal amounts of white in the axillaries or along the underwing coverts. Some sub-adult IV birds can appear almost identical to adults, whereas some Bald Eagles over five years of age can still have dark streaking on the head. Since individual eagles molt at slightly different rates, the plumage of older birds may not always correlate with their age in years. **Caution: Adult Bald Eagles can have a dark, smudgy tail tip, caused by mud stains, that resembles the dark tail tip of sub-adult birds.**

Bald Eagle

BE 01 - Bald Eagle, juvenile, sub-adult I (WA). Juveniles (left) are mostly dark underneath with white axillaries and varying amounts of white in underwing coverts. Sub-adult I birds (right) have white belly and uneven trailing edge on wings as a result of molt. Note that belly of juveniles fades by fall.

BE 02 - Bald Eagle, sub-adult II (MI). Sub-adult II birds can have either white (left) or dark belly (right). Sub-adult II birds may retain several juvenile (longer) secondaries, such as these, or may have an even trailing edge to wings. Difference in wing molt is usually the only way to tell sub-adult I from white-bellied sub-adult II birds.

BE 03 - Bald Eagle, sub-adult II, III (UT). Sub-adult II birds (left) begin to acquire whitish head but may still have mottled belly; note even trailing edge on wings. Sub-adult III (transition) birds (right) have mostly dark underwings and belly and mostly white head with dark eye-lines. Photo on right © Sherry Liguori

BE 04 - Bald Eagle, sub-adult IV. Similar to adults but with some dark feathers retained in tail and head, and possibly some white in body and wings (left, UT). Sub-adult IV birds can have nearly completely white head and tail but have fair amount of white in axillaries and secondaries (right, IL). Photo on right © Mike Lanzone

BE 05 - Bald Eagle, adult (WA). Adults are dark overall with brilliant white head and tail.

BE 06 - Bald Eagle, adult (WA). White head and tail of adult Bald Eagles can appear dark on overcast days. © Aaron Barna

BE 07 - Bald Eagle, adult (UT). Adults can appear headless when flying against a white backdrop such as snow or clouds. © Sherry Liguori

BE 08 - Bald Eagle, juvenile (UT). Juveniles are brown on top with contrasting black flight feathers.

BE 09 - Bald Eagle, sub-adult I, II (UT). Sub-adult I birds (left) are dark above with white back and uneven trailing edge on wings. Sub-adult II birds (right) may or may not have white on back but typically have pale crown and even trailing edge on wings.

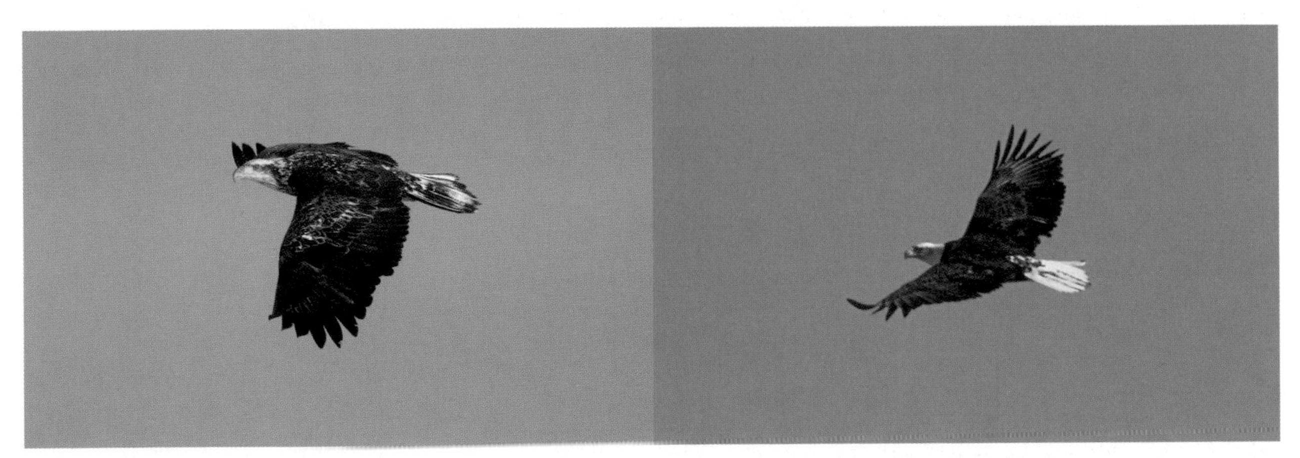

BE 10 - Bald Eagle, sub-adult III, IV (UT). Sub-adult III birds (left) are mostly dark on top with whitish head and dark eye-line. Tail varies from mostly dark to mostly white with dark tip. Sub-adult IV (transition) birds (right) are similar to adults but with some dark in tail and often behind eye. Photo on left © Sherry Liguori

BE 11 - Bald Eagle, adult (AK). Adults are all dark above with pure white head and tail. Note stain on tail tip from mud. © Gary Crandall

BE Pitfall 01. From eye level, white belly of sub-adult I birds (left, UT) can look like white axillaries of juvenile birds (right, MI). Note somewhat pale crown and uneven trailing edge on wings of sub-adult I bird. White belly of sub-adult I bird is shadowed.

BE Pitfall 02 (MI). Belly of juveniles (left) fades by spring and can appear whitish, similar to that of sub-adults (right). Note translucent "gaps" in wings and broader wing shape of juvenile, and retained juvenile secondaries of sub-adult II bird. Juveniles lack signs of molt.

BE Pitfall 03. Dark-bellied sub-adult II birds (left, WA) can appear similar to juveniles (right, MI). Note less extensive white in axillaries and underwing coverts of sub-adult II birds, and that translucent "windows" in wings have dark borders. Also note broader wings of juvenile.

BE Pitfall 04 (UT). Juveniles (left) can look similar to dark-backed sub-adult II birds (right). However, juveniles exhibit two-toned uppersides with brownish back and upperwings contrasting against blackish flight feathers. Note pale crown and uniformly dark upperwings of sub-adult II bird.

BE Pitfall 05 (NV). At a distance, transition **Bald Eagles** (left) can appear similar to immature **Golden Eagles** (right). Note whitish head of Bald Eagle and mottling along upperwing coverts of Golden Eagle. Dark tail tip of Golden Eagle is typically broader than that of transition Bald Eagle.

Golden Eagle (*Aquila chrysaetos*)

Golden Eagles are mostly dark with a golden nape. Juvenile and sub-adult birds have varying amounts of white in the tail and remiges, gradually becoming darker as they reach adulthood. Adults are completely dark with faint gray mottling in the tail and remiges. **Sub-adult remiges may have a prominent white base similar to those of juveniles; when they lack a white base, they are identical to adult remiges.** For this reason, the wings of sub-adult birds may appear identical to those of adults. By contrast, sub-adult tail feathers always have a white base. When seen well, the tail pattern is the most reliable trait to use for aging Golden Eagles, except on juveniles, where the upperwing pattern is the most reliable feature. In the field, age-related traits of Golden Eagles are often obscured, making classification to specific age impossible.

Juvenile, Sub-adult I, and Sub-adult II

Juvenile Golden Eagles typically have white patches at the base of the remiges that vary in size among individuals, from extensive to none at all. Birds with extensive white wing patches underneath often show small white patches on the upperwings. **The white patches in the wings of juveniles are often solid but may be divided by dark feathers and appear similar to the patches of older birds that have already experienced molt.** By contrast, older birds can have small, solid white patches in the wings; do not age Golden Eagles based on this trait alone. Juveniles have a white-based tail with a well-defined dark tip. The white in the tail is typically extensive, covering more than half the tail, but it can be fairly restricted in rare cases. The upperwings of juveniles are uniformly brown, but the upperwing coverts and golden nape typically fade by spring and appear paler than usual.

Sub-adult I Golden Eagles (1–2 years old) tend to retain most of their juvenile feathers and appear nearly identical to juveniles. On sub-adult I birds, primary 1 and secondary 1 are sometimes the only remiges replaced by sub-adult feathers. However, the trailing edge of the wings does not appear irregular since it tapers naturally where the primaries and secondaries meet. Sub-adult I birds may acquire several new secondaries, which can give the wings an uneven appearance. **Any eagle that is actively molting during fall and winter, or shows signs of previous molt, is not a juvenile. Be aware that certain aspects of molt are often obvious in photographs but can be difficult to observe in flight.** Since the inner primaries of juveniles can have faint gray banding like the primaries of older birds, assessing the presence of sub-adult primaries based on feather pattern is impossible.

On sub-adult I birds, the central (deck) tail feathers are sometimes the only tail feathers to be replaced by sub-adult feathers. Sub-adult tail feathers show a faint gray band below the dark tip but retain a broad white base, making them especially difficult to distinguish from juvenile feathers in flight. Varying amounts of upperwing coverts on sub-adult I Golden Eagles are replaced, with the new pale feathers forming a narrow, mottled "bar" along the upperwing. **Juveniles are the only age class that lacks a pale "upperwing bar," but their upperwing coverts may fade by spring, appearing similar to the upperwings on older birds.** The faded upperwings of juveniles in spring are relatively broad and even toned, lacking mottling.

The plumage of sub-adult II Golden Eagles (2–3 years old) comprises mostly sub-adult feathers, with some retained juvenile feathers. This makes sub-adult I and II Golden Eagles similar in appearance and often impossible to tell apart in flight. Rarely, the central tail feathers of sub-adult II birds may be replaced with adult feathers, causing the tail to look dark centered or "split." Sub-adult II birds often replace all of their juvenile secondaries, but they may retain a few juvenile secondaries on each wing, which project slightly beyond the trailing edge. Although this is uncommon, the retained juvenile feathers may fall even with the trailing edge of the wings when extremely worn.

Although a challenge, assessing molt in flight can prove extremely useful in aging Golden Eagles. For example, sub-adult I and II birds that are actively molting in fall can be distinguished from each other by the difference in their primary molt pattern. **On sub-adult II birds, primary molt occurs toward the**

outer end of the wings; sub-adult I birds replace only inner primaries.

Since sub-adult II Golden Eagles typically have a complete set of sub-adult or adult-like remiges, they appear two-toned on the underwings, similar to older Golden Eagles. This two-toned appearance is most visible when the wings are illuminated by pale ground cover. The flight feathers of Golden Eagles may appear almost whitish when illuminated by snow cover, and the dark trailing edge of the wings, which juvenile and sub-adult I birds lack, becomes more apparent.

Sub-adult III

Sub-adult III Golden Eagles (3–4 years old) typically show a mix of adult and sub-adult tail feathers. Because of the retained sub-adult tail feathers, the tail will show white patches on each side. When seen from below, this "split-tailed" appearance is difficult to observe when the tail is folded, and sub-adult III birds appear much like sub-adult IV and adult birds. **Observing the tail pattern from above is often necessary to accurately age sub-adult Golden Eagles.** Sub-adult III birds typically have a complete set of adult-like remiges, but in rare cases they may retain a few juvenile secondaries. When this occurs, sub-adult II and III Golden Eagles are nearly impossible to tell apart in flight. Because of molt patterns, sub-adult II birds typically appear more ragged than other Golden Eagles.

Sub-adult IV and Adult

Sub-adult IV birds (4–5 years old), or birds immediately preceding adulthood, often possess a complete set of adult remiges and look identical to adults, which lack white patches in the tail and wings. At times they may retain one or two sub-adult secondaries that may or may not show a white base. **The presence of sub-adult tail feathers (typically next-to-outer feathers) on sub-adult IV Golden Eagles is often the only way to tell them from adults.** The white base of sub-adult tail feathers is most apparent when the tail is spread. Because of distance or poor visibility, however, it is often difficult to see the retained sub-adult tail feathers of sub-adult IV Golden Eagles even with the tail is spread. **When the tail is folded, it is often impossible to specifically age Golden Eagles from underneath, especially when seen gliding overhead in typical migratory fashion.** This is especially true of sub-adult IV Golden Eagles, which can only be aged from above when the tail is folded.

Golden Eagle

GE 01 - Golden Eagle, juvenile. White wing patches vary in size among juveniles. Note bird on right (UT) has a somewhat grayish tail base; however, it spans entire width of tail as in all juveniles. Juveniles typically begin their first molt in April, like bird on right (note primary 1 growing in).

GE 02 - Golden Eagle, sub-adult I, II. With neat white tail base, this sub-adult I bird (left, NV) appears juvenile-like. Note "gap" in wings caused by molt. White tail base of sub-adult II birds (right, WY) is less neat and uneven. Note outer primary molt as opposed to inner primary molt of sub-adult I bird. Photo on right © Sherry Liguori

GE 03 - Golden Eagle, sub-adult II, III. Tail of this sub-adult II bird (left, NM) is obscured. However, aging it is possible by few retained juvenile (longer) secondaries. Sub-adult III birds (right, NV) have dark outer and central tail feathers with white base otherwise. Photo on right © Chris Neri

GE 04 - Golden Eagle, sub-adult IV (UT). Sub-adult IV birds can be aged from below when tail is spread (left); note small white patches in next-to-outer sub-adult tail feathers. Distinguishing sub-adult IV from adult when tail is folded (right) is impossible from below (same bird as on left). Both photos © Sherry Liguori

GE 05 - Golden Eagle, adult. Adults have a complete set of adult remiges and tail feathers, which are dark with faint, grayish bands (left, NV). In spring and summer, molting adults can show white areas in underwing coverts that appear similar to white wing patches of immature birds (right, UT).

GE 06 - Golden Eagle, juvenile (NV). Juveniles are the only age class that lacks a tawny "upperwing bar" (left). Some juveniles, especially birds with large white wing patches underneath, exhibit small white upperwing patches (right). Note white-based tails. Photo on left © Chris Neri

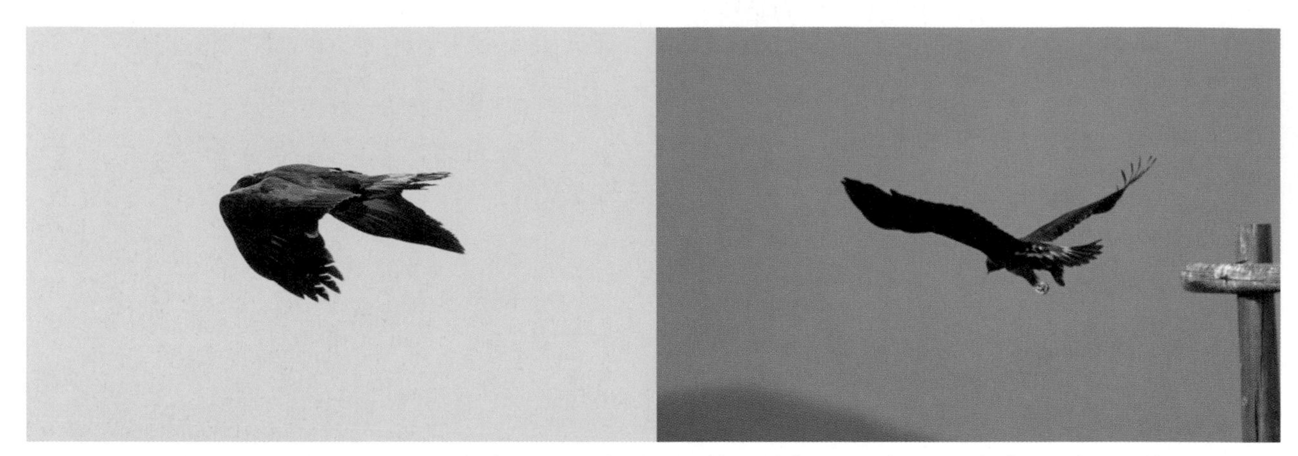

GE 07 - Golden Eagle, juvenile. Some juveniles have grayish white tail base (left, MI). White in tail of juvenile on right (NV) is somewhat inconspicuous and limited to inner third of tail. Note lack of tawny "upperwing bar." Photo on left © Michael Shupe

GE 08 - Golden Eagle, sub-adult I, II. In flight, sub-adult I (left, UT) and sub-adult II (right, NV) birds can look identical. Note white-based tails, similar to juveniles, and tawny "upperwing bars." Almost all juvenile primaries (faded) of sub-adult II are replaced, but this is often impossible to judge in flight.

GE 09 - Golden Eagle, sub-adult III, IV. Note mix of adult (dark) and sub-adult (white-based) tail feathers on sub-adult III bird (left, CA). Sub-adult IV birds (right, UT) typically retain only next-to-outer sub-adult tail feathers. Photo on right © Sherry Liguori

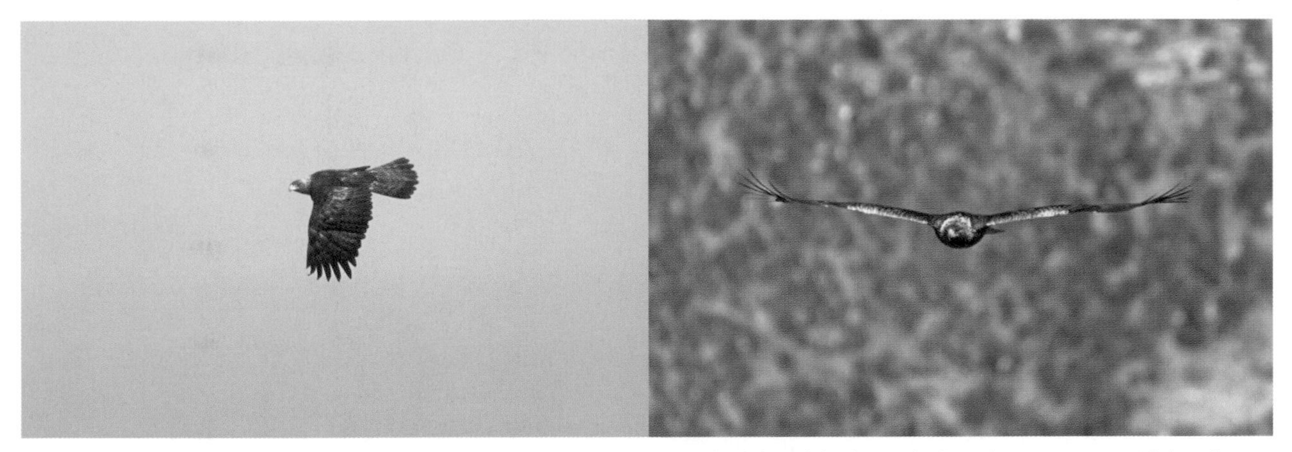

GE 10 - Golden Eagle, adult. Adults (left, UT) show a complete set of adult tail feathers (dark with narrow, grayish bands). From head-on (right, NV), Golden Eagles are dark with golden head and pale leading edge on wings. In fall, adults and sub-adults tend to have paler head than juveniles.

GE Pitfall 01. Dark morph **Red-tailed Hawks** (juvenile, left, NV) are similar to **Golden Eagles** (adult, right, UT). However, Golden Eagles show dark flight feathers and relatively longer wings compared with Red-tailed Hawks. Photo on left © Chris Neri

EAGLE PITFALLS

Plumage

When observed from a side angle, Bald Eagles can be tricky to age correctly. Since the belly is often obscured from this angle, **the axillaries of juveniles can be mistaken for the belly, causing juvenile Bald Eagles to look like white-bellied Bald Eagles.** Remember that white-bellied Bald Eagles have white on the back and show obvious wing molt. Sub-adult II Bald Eagles with a dark belly and dark head can appear similar to juveniles from this angle as well. Most sub-adult II birds, however, have limited white feathers on the axillaries and underwing coverts, show a pale crown, and often have white on the back.

At times, the effects of molt make it more difficult to classify eagles to age. During spring and summer when adult Golden Eagles molt their underwing coverts, the white base of the remiges normally covered by dark underwing linings becomes exposed. When this occurs, adults show small white patches in the wings similar to those of immature birds. However, **these white patches on adults appear irregular and are limited to the wing linings.** Be aware that on bright days the sun's glare can cause the grayish tail bands on the topside of adult Golden Eagles to appear whitish, similar to the tail of immature birds. Also, the golden undertail coverts of Golden Eagles can appear somewhat whitish when well illuminated. Regardless, **the tail and undertail coverts of adult Golden Eagles never look brilliant white like the tail base of immature birds.**

Similar Species

Since both Ospreys and juvenile Bald Eagles exhibit contrasting white and black undersides, they can sometimes be confused with light morph Swainson's Hawks. Ospreys, however, show a bolder black and white pattern on the underside and are much lankier in shape than Swainson's Hawks. The white areas on Juvenile Bald Eagles are limited to the axillaries and underwing linings, and they show a less contrasting two-toned underside than light morph Swainson's Hawks. Juvenile Bald Eagles are also broader overall in shape than Swainson's Hawks. Both Ospreys and eagles lack the pointed wing tips that Swainson's Hawks exhibit.

With long, broad wings often held in a dihedral, dark morph Red-tailed, Ferruginous, and Rough-legged Hawks resemble Turkey Vultures and Golden Eagles. Ferruginous and Rough-legged Hawks sometimes "wobble" in flight like Turkey Vultures, but their wings are narrower and their head is larger than that of vultures. By contrast, Golden Eagles have a larger head and longer wings than buteos. The wings of Turkey Vultures and Golden Eagles may exhibit silvery flight feathers, but their wings are never as contrasting underneath as those of dark buteos, especially Ferruginous Hawks, which exhibit whitish flight feathers. Adult Golden Eagles can show a dark terminal band on the wings when well lit, but since Golden Eagles have dark flight feathers, the terminal band is typically not as distinct as that of buteos. When soaring, eagles turn in slow, wide circles compared to buteos. When flapping, the wing beats of Golden Eagles are much more labored than those of buteos.

FLIGHT STYLE

Large birds typically appear slow moving and steady in flight. Turkey Vultures, however, often fly in a wobbly manner, constantly swaying from side to side. **This manner of flight makes Turkey Vultures distinctive among the larger raptors.** Black Vultures are steady in flight but can exhibit unsteadiness in high winds. Vultures, Ospreys, and eagles soar in wide, lazy circles, climbing on thermals at a slower rate than smaller raptors. Ospreys and Turkey Vultures are more buoyant than Black Vultures and eagles and appear to gain lift and stay afloat easier. In all postures, Ospreys hold their wings bowed downward. Turkey Vultures fly with wings held in a strong or modified dihedral. Black Vultures soar on flat wings or with wings held in a slight dihedral, but they glide with a slight modified dihedral. Bald Eagles typically hold their wings flat or slightly bowed downward, whereas Golden Eagles usually show a shallow to moderate dihedral in a soar and slightly bowed wings in a glide.

Wing Beat

Each of the large raptors flaps in a distinct manner. **Ospreys exhibit stiff, shallow, labored wing beats compared to Turkey Vultures and eagles.** When Ospreys beat their wings forcefully to maintain altitude, they seem to struggle, as their head and chest appear to bob up and down. When flapping, Black Vultures appear labored as well; however, **the wing beats of Black Vultures are quick, shallow, and snappy compared with other large raptors.**

Turkey Vultures exhibit lofty, easy wing beats that end abruptly at the top of an upstroke. Coincidentally, all raptors that fly with a dihedral end their wing beats abruptly on an upstroke. This is an excellent way to distinguish the wing beats of Bald and Golden Eagles. **Golden Eagles display shallow, easy wing beats that end abruptly at the top of an upstroke, forming a dihedral. Bald Eagles exhibit stiff, lofty wing beats that end on an upstroke but immediately level out.** Adult Bald Eagles are more likely to end their wing beats abruptly on a downward flap, whereas juveniles tend to settle their wings in a level manner. During light winds at ridge sites, Golden Eagles often exhibit shallow, "wristy," intermittent flaps to maintain altitude. These periodic wing beats also end abruptly on an upstroke.

SOARING

Vultures, Ospreys, and eagles are distinctive in a soar and can often be identified at great distances with relative ease; however, telling Bald and Golden Eagles apart can be more difficult. Luckily, key plumage traits of eagles, such as the white wing patches on immature Golden Eagles and the white belly of immature Bald Eagles, are often obvious even in poor light. Turkey Vultures have long, broad wings that lack a noticeable taper and long tails that are typically wedge shaped at the tip. Black Vultures have short, broad, rounded wings with squared-off hands; the tail is extremely short and square tipped. **Black and Turkey Vultures look particularly small headed in flight.** Turkey Vultures soar with an exaggerated dihedral. Black Vultures soar on flat wings or with a slight dihedral.

Black Vultures can look similar to juvenile Bald Eagles, especially when soaring at eye level. Both soar on relatively flat, broad wings and can appear completely dark in the field, but the wings and especially the tail of Black Vultures are stockier than those of Bald Eagles. The wings of Black Vultures arch forward in a soar, unlike the wings of Turkey Vultures and Bald Eagles, which are held straight out in a soar. **When banking, juvenile Bald Eagles show a two-toned pattern on the upperside that Black Vultures never show.**

Ospreys are as distinctive in shape as they are in plumage. They have extremely long, narrow wings that bulge slightly at the base and taper slightly toward the tips. The bold black and white pattern on the underside of Ospreys is obvious overhead, even in poor light. Adult Bald Eagles, which possess a white head similar to Ospreys, are sometimes confused with Ospreys. However, **Bald Eagles always exhibit broader, more straight-cut wings and a stockier body than Ospreys.** Bald Eagles hold their wings flat in a soar, whereas the wings of Ospreys droop slightly.

Bald and Golden Eagles exhibit extremely long, broad wings in a soar, but Bald Eagles have slightly lankier wings than Golden Eagles. **The wings of Golden Eagles taper, or "pinch in," toward the body along the leading and trailing edges of the wings, as in many buteos.** The wings of Bald Eagles are straight along the leading edge and taper only slightly toward the body along their trailing edge. The head of Bald Eagles protrudes farther past the leading edge of the wings than in Golden Eagles; this is most evident when seen soaring overhead. Golden Eagles exhibit a moderate dihedral in a soar, which is more pronounced than the shallow dihedral sometimes displayed by Bald Eagles. **Remember, immature Bald Eagles show small translucent windows at the tips of the inner primaries; Golden Eagles do not.** Although the wings of juvenile eagles are slightly broader than those of adults, this trait is difficult to judge without practice.

HEAD-ON

Turkey Vultures are the most distinctive of the large raptors when approaching at eye level. **Turkey Vultures typically hold their wings in a modified dihe-**

dral, whereas Ospreys and eagles hold their wings bowed downward. The primaries, or fingers, of Turkey Vultures are often distended and visible, whereas the hands of Ospreys and eagles are typically drawn in when gliding. Black Vultures exhibit short wings compared to Turkey Vultures, Ospreys, and eagles. Black Vultures typically hold their wings somewhat flat when approaching head-on but may show a slight modified dihedral. **Ospreys exhibit long, narrow wings that rise at the shoulders and droop sharply at the wrists.** Ospreys appear particularly narrow at the chest compared to vultures and eagles. Whereas vultures appear all black, the white head or chest of Ospreys is usually obvious from a head-on view.

When viewed approaching at eye level, Bald and Golden Eagles are very similar in shape. **The wings of Golden Eagles are raised slightly at the shoulders and bowed down at the hands similar to the posture of Red-tailed Hawks; the wings of Bald Eagles bow more evenly and smoothly throughout their length.** This gives the wings of Golden Eagles a more angular appearance than those of Bald Eagles. In a shallow glide, Golden Eagles may show a slight dihedral; Bald Eagles can exhibit a slight modified dihedral but only during high winds. **When viewed head-on, Golden Eagles exhibit a pale leading edge to the wings; Bald Eagles do not.** Be aware that Golden Eagles can appear white-headed when the golden nape is illuminated by direct sunlight.

GLIDING OVERHEAD

Vultures, Ospreys, and eagles are easy to identify from other raptors when gliding overhead; however, telling the shapes of Bald and Golden Eagles apart can be tricky. **Black Vultures possess extremely stocky wings with short hands that project slightly beyond the wings.** In a glide overhead, the short, narrow, square-tipped tail of Black Vultures is obvious. Turkey Vultures are more similar to eagles than to Black Vultures and Ospreys but show less tapered wings than eagles. Vultures also have a small head compared to eagles. **Ospreys have extremely long, narrow wings that form a distinct M shape when drawn in.** The body and tail are also relatively narrow on Ospreys compared with vultures and eagles. Bald Eagles exhibit a shape similar to Golden Eagles. However, the head of Bald Eagles appears larger and the wings are somewhat straight cut. **Golden Eagles have slightly narrower, more tapered hands than Bald Eagles, and they curve smoothly when swept back.**

WING-ON/GOING AWAY

From a side angle, Ospreys show extremely long, narrow, bowed wings. They are chesty with a slim belly compared to vultures and eagles, which exhibit a broader body and wings. From above, transition Bald Eagles with a white head and dark eye-line are the most similar of the raptors to Ospreys. Although transition Bald Eagles have a mostly or all white tail, the tail of Ospreys is mostly dark above, even when fanned.

From wing-on, vultures lack a taper to their wings, and their head appears narrow. **Turkey Vultures hold their wings in a modified dihedral that is accentuated in high winds.** Black Vultures hold their wings somewhat flat but may show a slightly drooped, modified dihedral. Vultures appear black on top and lack fading or mottling on the upperwings, which most eagles exhibit.

The wings of Bald Eagles, unlike those of Golden Eagles, appear to lack any bulge, and they taper to a point along the leading edge. Remember that the wings of Golden Eagles are raised slightly at the shoulders and bowed down at the hands; the wings of Bald Eagles bow gently throughout their length, appearing somewhat flatter than those of Golden Eagles. **The head of Bald Eagles protrudes farther beyond the leading edge of the wings than that of Golden Eagles**, similar to the difference in head projection between Sharp-shinned Hawks and the larger-headed Cooper's Hawks. **Golden Eagles never show a distinctly two-toned upperside or white back like immature Bald Eagles do.**

When headed away, Ospreys exhibit narrow, bowed wings which appear extremely long at the hands. The body and tail are also long and narrow. The wings of vultures are stocky at the base with long hands but are held flat or raised at the shoulders, and they are broader overall than the wings of Ospreys. Black Vultures exhibit noticeably shorter wings and tail than Turkey Vultures when headed away. The wings of eagles are long compared to those of vultures when viewed headed away, and their body is broad, or "heavy." **The wings of Golden Eagles show an obvious taper toward the body, whereas the wings of Bald Eagles are straighter across the trailing edge.**

Osprey, Eagle, and Vulture Shapes

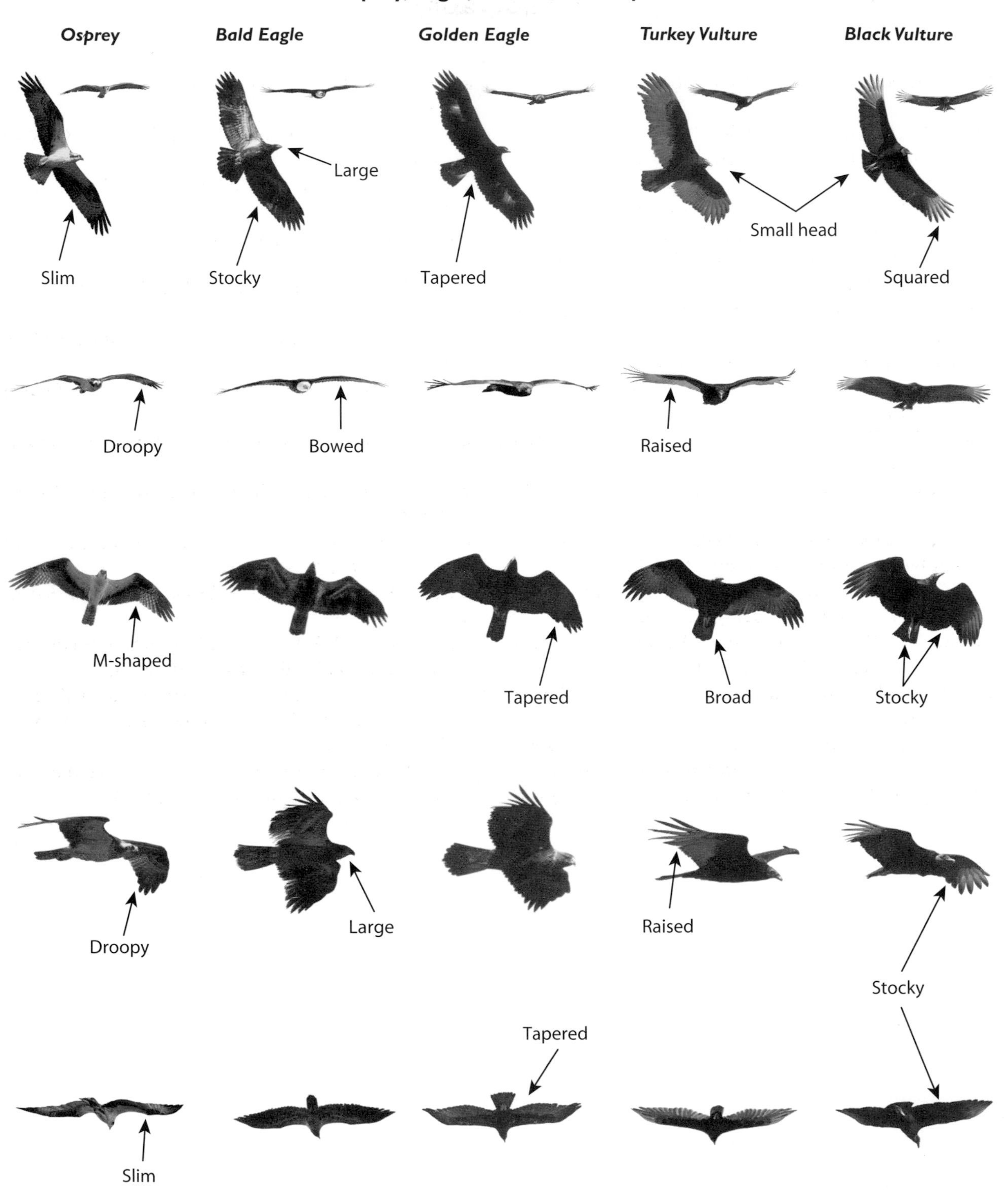

Bibliography

Clark, W. S. 2001. Aging Bald Eagles. *Birding* 33(1): 19–28.

Clark, W. S., and B. K. Wheeler. 1997. *A Photographic Guide to North American Raptors.* Academic Press (San Diego, Calif.).

Clark, W. S., and B. K. Wheeler. 2001. *Hawks of North America.* 2d ed. Houghton Mifflin (Boston).

Dunne, P., D. Sibley, and C. Sutton. 1988. *Hawks in Flight.* Houghton Mifflin (Boston).

Griggs, J. L., and J. Liguori. 1998. *All the Birds of Prey.* Harper Collins (New York).

Heintzelman, D. S. 1979. *A Guide to Hawk Watching in North America.* Pennsylvania State University Press (University Park).

Liguori, J. 1997. Answers to August Photo Quiz. *Birding* 29(5): 411–14.

Liguori, J. 2000. Enraptured by Raptors. *WildBird* 14(8): 30–33.

Liguori, J. 2000. *Hawks of Whitefish Point.* Whitefish Point Bird Observatory (Paradise, Mich.).

Liguori, J. 2000. Identification Review: Sharp-shinned and Cooper's Hawks. *Birding* 32(5): 428–33.

Liguori, J. 2001. Choosing Optics—A Matter of Personal Preference. *Raptor Watch* 15(2): 12.

Liguori, J. 2001. Pitfalls in Osprey Identification. *Raptor Watch* 15(3): 8.

Liguori, J. 2001. Pitfalls of Classifying Light Morph Red-tailed Hawks to Subspecies. *Birding* 33(5): 436–46.

Liguori, J. 2001. Revered Raptors. *WildBird* 15(4): 26–33.

Liguori, J. 2002. Identifying Silhouettes. *Raptor Watch* 16(3): 5.

Liguori, J. 2002. Raptor Migration Studies in the Goshute Mountains, Nevada. *Birding* 34(5): 444–49.

Liguori, J. 2002. Raptors in Abundance. *WildBird* 16(4): 30–33.

Liguori, J. 2003. Birding Whitefish Point. *Bird Watcher's Digest* 25(3): 58–61.

Liguori, J. 2004. August Photo Quiz Answers. *Birding* 36(5): 520–22.

Liguori, J. 2004. Dark Red-tailed Hawks. *Birding* 36(5): 501–06.

Liguori, J. 2004. How to Age Golden Eagles. *Birding* 36(3): 278–83.

Liguori, J. 2004. Species Profile: American Kestrel. *Raptor Watch* 18(1): 8–10.

Liguori, J., and S. Liguori. 2004. Going My Way? *WildBird* 18(2): 46–51.

Liguori, J., and S. Liguori. 2004. Species Profile: Red-tailed Hawk. *WildBird* 18(4): 46–9.

Sibley, D. A. 2000. *The Sibley Guide to Birds.* Alfred A. Knopf (New York).

Sutton, C., and P. T. Sutton. 1996. *How to Spot Hawks and Eagles.* Chapters Publishing Ltd. (Shelburne).

Walton, R. K., and G. Dodge. 1997. *Hawk Watch: A Video Guide to Eastern Raptors.* Brown Bag Productions (Hillsborough, N.C.).

Wheeler, B. K. 2003. *Raptors of Western North America.* Princeton University Press (Princeton, N.J.).

Index

English names are written in roman type; scientific names are written in italic type.

Accipiter
- *cooperii*, 15
- *gentilis*, 15
- *striatus*, 15

Aquila chrysaetos, 120

Buteo
- *albonotatus*, 109
- *jamaicensis*, 53
- *lagopus*, 72
- *lineatus*, 43
- *platypterus*, 46
- *regalis*, 67
- *swainsoni*, 49

Cathartes aura, 109
Circus cyaneus, 31
Coragyps atratus, 108

Eagle
- Bald, 107, 114–119, 126–130
- Golden, 36, 76, 107, 119–130

Falco
- *columbarius*, 88
- *mexicanus*, 98
- *peregrinus*, 94
- *rusticolus*, 99
- *sparverius*, 86

Falcon
- Peregrine, 85, 94–97, 100, 102–105
- Prairie, 85–86, 98–99, 102–105

Goshawk, Northern, 15–30, 97
Gyrfalcon, 99–100

Haliaeetus leucocephalus, 114
Harrier, Northern, 31–39, 97
Hawk
- Broad-winged, 41–42, 46–48, 78–84, 97
- Cooper's, 15–30, 36
- Ferruginous, 41–42, 67–71, 76, 78–84
- Red-shouldered, 41–45, 48, 78–84
- Red-tailed, 41–42, 53–66, 70, 78–84, 125
- Red-tailed (Krider's), 53–54, 57–58, 70
- Rough-legged, 41–42, 71–84
- Sharp-shinned, 15–30, 93
- Swainson's, 41–42, 49–52, 78–84
- Zone-tailed, 109–111

Ictinia mississippiensis, 100

Kestrel, American, 85–88, 92–93, 102–105
Kite, Mississippi, 100–102

Merlin, 85, 88–93, 102–105

Osprey, 107, 112–113, 126–130

Pandion haliaetus, 112

Vulture
- Black, 107–109, 126–130
- Turkey, 107, 109–111, 126–130